F. Rufenacht Walters

Sanatoria for Consumptives in Various Parts of the World

France, Germany, Norway, Russia, Switzerland, the United States and the British

Possessions. A critical and detailed description together with an exposition of the

open-air or hygienic treat

F. Rufenacht Walters

Sanatoria for Consumptives in Various Parts of the World
France, Germany, Norway, Russia, Switzerland, the United States and the British Possessions.
A critical and detailed description together with an exposition of the open-air or hygienic
treat

ISBN/EAN: 9783337191993

Printed in Europe, USA, Canada, Australia, Japan

Cover: Foto ©berggeist007 / pixelio.de

More available books at **www.hansebooks.com**

SANATORIA FOR CONSUMPTIVES

ABERDEEN UNIVERSITY PRESS

SANATORIA FOR CONSUMPTIVES

IN VARIOUS PARTS OF THE WORLD

(FRANCE, GERMANY, NORWAY, RUSSIA, SWITZERLAND, THE UNITED STATES AND
THE BRITISH POSSESSIONS)

A CRITICAL AND DETAILED DESCRIPTION

TOGETHER WITH AN EXPOSITION OF

THE OPEN-AIR OR HYGIENIC TREATMENT OF PHTHISIS

BY

F. RUFENACHT WALTERS, M.D., M.R.C.P.

FELLOW OF THE ROYAL COLLEGE OF SURGEONS ; PHYSICIAN TO THE NORTH LONDON
HOSPITAL FOR CONSUMPTION AND DISEASES OF THE CHEST

WITH AN INTRODUCTION

BY

SIR RICHARD DOUGLAS POWELL, BART., M.D., F.R.C.P.

LONDON

SWAN SONNENSCHEIN & CO., LIM.

PATERNOSTER SQUARE

1899

PREFACE.

By Sir Richard Douglas Powell.

The present little work of Dr. Walters, which is based upon careful investigation and expert knowledge, brings before us in a purposively arranged manner the most modern methods of hygienic treatment of consumption. The chief object of the author is to advocate the establishment in this country of public institutions of the convalescent hospital class, and of sanatoria designed for the reception of cases of tuberculosis from amongst those who are more or less well-to-do.

Such institutions already exist on the continent and in America; and there is no doubt whatever that there are numbers of patients of both kinds sufficient to keep such sanatoria very usefully occupied. Nor can any thoughtful person doubt the value of such sanatoria, and the boon that they would be to a large number of people in this country, who under conditions of prolonged illness have no (or at least no adequate) home accommodation available.

Their usefulness, however, in my judgment, extends far beyond the immediate purpose for which they are proposed. Lessons in self-management are learned by those who sojourn for a time in such sanatoria; habits of self-discipline and attention to hygienic laws are acquired which are of much importance to those afflicted with consumption, and which have a favourable influence upon prophylaxis; and these persons when they pass again into the general community become centres of instruction in domestic hygiene.

It is only during certain periods or phases of this usually prolonged malady that sanatoria can ever be available; and many patients are able to carry out, and will prefer, home treatment. For consumption is a disease that can

influence over the incidence and progress of consumption ;
but the value of particular climates has been much exaggerated, and that of fresh air *per se* has been insufficiently
realised in many quarters.

To a large extent the circumstances which predispose to
an attack of consumption also hinder recovery from it ; so
that in order to treat the disease with success they must all
be taken into account. Some physicians may be sceptical
as to the value of particular remedies in the treatment of
consumption ; but all are agreed as to the supreme importance of hygienic measures.

The treatment which has led to the best results—that
carried out in many sanatoria and health resorts—is based
on a minute regulation of daily life in all its hygienic
bearings, together with such other measures as may be
indicated by the needs of each patient. Experience has
shown that much may be accomplished by an out-of-door
life in a suitable neighbourhood, scrupulously careful
ventilation of rooms, systematic regulation of the amount
of rest and exercise according to the passing state of the
patient, attention to the skin and digestive organs, a
carefully chosen dietary, and other largely hygienic
measures. A far-reaching programme of this kind can
only be carried out in the first instance where the physician
is able to exercise almost constant supervision and control ;
at certain stages it cannot easily be carried out excepting
in a sanatorium specially arranged and devoted to this end.

Up to recent times England had probably done more to
reduce phthisical mortality than any other country ; but
she is now being outstripped in the race by several others,
notably by Germany, in which several dozen sanatoria
have been erected during the last few years, and much
useful work done in other directions.

Public opinion on the continent, both lay and medical, is
overwhelmingly in favour of the sanatorium system. In
our own country many physicians—an increasing number
—are also favourable to it, and recommend its adoption in

this country with suitable modifications. It has been tried here and there on a small scale in England, and is more or less the basis of treatment at the Ventnor Consumption Hospital; and although up to the present time no sanatorium of any size has been established where the treatment can be completely carried out in a satisfactory manner, yet the results already obtained with imperfect means fully justify a favourable opinion.

The present volume is intended to present a general review of the subject, and a detailed description of existing sanatoria, including their climate, situation, construction and arrangement, their management, methods of treatment, and results. For this purpose a large amount of original information has been obtained from medical officers and others attached to the various institutions, which has been carefully compared with published descriptions and in many cases checked by personal visits. The literature of the subject is already voluminous, and in many different languages; but no approximately complete account has yet been published, the best works—written respectively in French and in Hungarian—being already out of date, while most of the existing sanatoria have not yet been described in any English book or periodical.

Continental progress in sanatorium treatment has been so rapid of late that omissions and inaccuracies are inevitable in a book of this nature, although they have as far as possible been avoided. Two additional sanatoria for consumptives are being erected in the United States of America, one in connection with a large asylum for the insane at Chicago, the other in connection with the home for incurables at the same place (Klebs). In Germany several of the sanatoria described as in course of erection have now been completed, while others are being projected at Fürth and Liegnitz. The Berka sanatorium in Thuringia, which has just been opened, is to receive the name of *Sophien-Heilstätte*. Prof. Korányi's Sanatorium Society at Buda-Pesth has decided to call its next sanatorium for con-

sumptives *Elisabeth-Sanatorium für Lungenkranke*, in honour
of the late Empress.   It has been decided in Hungary to
permit workmen who are obviously consumptive to work
only in special work places, and to enforce stricter hygienic
measures in factories generally.   In Russia a children's
sanatorium has been erected not far from Kief, and
another is to be erected for children of the poor.   In
Switzerland the Zurich sanatorium has just been opened.
Mention should have been made of the schools at Davos
for delicate children, and of the recent introduction of an
electric engine on the line from Landquart to Davos.   In
Holland a subscription of £25,000 was raised as a gift to
the Queen Regent in celebration of the coronation festivi-
ties ; the Queen Regent has however directed the sum to
be employed in the erection of a sanatorium for consumptives, and has further given her property " Oranje-Nassau
Oord " near Borkum for the purpose.

In preparing this work I have received much assistance
from the works of Kuthy, Hohe, Léon Petit, v. Jaruntowsky (trans. Beale) and Möller ; from the reports of Liebe
in the *Hygienische Rundschau*, and from various numbers
of *Das Rothe Kreuz* and *Heilstätten-Korrespondenz*.   I have
also to acknowledge the kind co-operation of a large
number of medical men and others in various parts of the
world, who have furnished me with most valuable information.   Among these Dr. Knopf, Dr. Stubbert, Dr. Trudeau
and Dr. Solly in the United States ; Dr. Derecq, Dr. Léon
Petit and Dr. Georges Petit in France; Dr. Römpler, Dr.
Weicker, Dr. Dettweiler, Dr. Meissen, Dr. Hess, Dr.
Gebhard and Dr. Wolff-Immermann in Germany ; Dr.
Turban, Dr. Jacobi, Dr. Huggard and Dr. Kündig in
Switzerland, call for special mention ; as well as those
who kindly furnished the originals of the various figures
in the text.   I add a list of those who kindly furnished
me with information, to whom my best thanks are due.

In America : Dr. Angney, Dr. Bowditch, Dr. Bell, Dr.
Bailey, Dr. Cauldwell, Dr. Davidson, Dr. Fisk, Dr. Fränkel,

Dr. Klebs, Dr. Knopf, Dr. Langdon, Dr. Miller, the Rev. F. W. Oakes, Dr. v. Ruck, Dr. Reynolds, Dr. Riley, Dr. Stubbert, Dr. Solly, Dr. Trudeau, and the Matron of the De Peyster Hospital.

In Austria-Hungary : Dr. Alex. Ritter v. Weissmayr, Prof. Korányi and Dr. Desider Kuthy.

In France: Dr. Armaingaud, Dr. S. Bernheim, Dr. Chaumier, Dr. Crouzet, Dr. Derecq, Dr. Giresse, Dr. Léon Petit, Dr. Georges Petit and Dr. Sabourin.

In Germany: Dr. Achtermann, Dr. Andrae, Dr. Baudach, Dr. Brecke, Dr. Boyé, the Directors of the Baden Anilin Fabrik, Dr. Besold, Prof. Dr. Dettweiler, Dr. Gebser, Dr. Gebhard, Dr. Gerhardt, Dr. Hess, Dr. Hauffe, Dr. Jacubasch, Dr. Krebs, Dr. Köhler, Dr. Kremser, Prof. Dr. Kobert, Dr. A. Koch, Dr. Ladendorf, Prof. E. v. Leyden, Dr. Liebe, Dr. Lehrecke, Dr. Lahmann, Dr. Meissen, Dr. May, Dr. Michaelis, Dr. Nahm, Dr. Ott, Dr. Pannwitz, Dr. Reuter, Dr. Römpler, Dr. Stauffer, Dr. Wolff-Immermann, Dr. Weibel, Dr. Weicker and Dr. Walther.

In Norway : Dr. Kaurin, Dr. Sömme.

In Russia : Dr. Treu, Dr. Unterberger, Dr. F. Weber.

In Switzerland: Dr. Exchaquet, Dr. Huggard, Dr. Glaser, Dr. Jacobi, Dr. Kündig, Dr. Philippi, Dr. Stephani, Dr. Schnöller, Dr. Turban and Dr. Wolff.

In Holland : Dr. Lismann.

In Australia : Dr. Kyngdon.

In England : Dr. Burton Fanning, Dr. Coghill, Dr. Johns, Dr. Pott, Dr. Pruen, Dr. M. Williams, Dr. Jane Walker.

If any of my informants have not been mentioned in this list, I must beg them to believe the omission to be quite unintentional on my part.

I am also particularly indebted to Sir Richard Douglas Powell for his kindness in contributing an Introduction to this book, as well as for very welcome encouragement and advice.

F. R. WALTERS.

December, 1898.

CONTENTS.

CONTENTS. XV

CONTENTS. xvii

CHAPTER XLI. SANATORIA IN NORWAY AND SWEDEN.

Table of Norwegian Sanatoria, with Beds and Altitudes—Tonsaassen
Sanatorium—The Gausdal Sanatorium—The Reknaes Consumption
Hospital—Other Sanatoria for the Poor in Norway . *page* 267

CHAPTER XLII. RUSSIAN SANATORIA.

Existing and Projected Sanatoria—Table of Sanatoria, with Beds and
Altitudes—Lindheim Sanatorium—Quisisana Sanatorium at Yalta—
The Sanatoria at Halila: the Alexander Sanatorium; the Maria
Sanatorium; the Nikolaj Sanatorium—The Taitzi Sanatorium –The
Military House Sanatoria at Zarskoje Selo and at Wola *page* 272

CHAPTER XLIII. SWISS SANATORIA FOR PAYING PATIENTS.

Table of Swiss Sanatoria, with Beds and Altitudes—The Arosa Sanatorium
—Davos—Dr. Turban's Sanatorium—The New Davos Sanatorium—
The Maison de Diaconesses—The Villa Pravignan—The Leysin Sanatorium—The Montana Sanatorium—The Weissenburg Kurhaus
page 283

CHAPTER XLIV. SWISS SANATORIA FOR POORER PATIENTS.

The Basel Sanatorium at Davos—The Dutch Sanatorium at Davos—The
Davos Invalids' Home—The Davos Benevolent Society—The St.
Moritz Aid Fund—The Heiligenschwendi Sanatorium—The Zurich
Sanatorium — The Sanatorium of Braunwald — The Aegeri Sana-
torium . . . . *page* 299

CHAPTER XLV. PROJECTED SANATORIA IN OTHER FOREIGN COUNTRIES.

The Sanatorium Movement in Belgium—Bulgaria—Denmark—Egypt—
Holland—Italy—Japan—Portugal—Roumania—Spain . *page* 311

CHAPTER XLVI. BRITISH COLONIAL SANATORIA.

Sanatoria in Australia and New Zealand—Canadian Sanatoria—The
Sanatorium in Cape Colony . . . . . . *page* 315

CHAPTER XLVII. BRITISH INSTITUTIONS FOR POORER CONSUMPTIVES.

London Institutions for Consumptives—Table showing Date of Foundation
and Number of Beds—Provincial Chest Hospitals and Sanatoria—
Table showing Date of Erection and Number of Beds—Homes for
Advanced Consumption—Table showing Date of Erection and Num-
ber of Beds—The Ventnor Consumption Hospital—The National
Sanatorium for Consumption at Bournemouth—The Cromer Experi-
mental Sanatorium—The Downham Sanatorium for the Poor—The
Victoria Hospital for Consumption at Craigleith—The Bridge of Weir
Hospital for Consumption—The National Hospital for Consumption
for Ireland—The Forster Green Hospital . . . *page* 318

CHAPTER XLVIII. BRITISH SANATORIA FOR PAYING CONSUMPTIVES.

Table of such Sanatoria, with Altitude and Number of Beds—The Bournemouth Sanatoria—Dr. Pott's Sanatorium—Dr. Johns' Sanatorium—
The Cotteswold Sanatorium near Cheltenham—The Downham Sanatoria—The Mundesley Sanatorium . . . . . *page* 335

b

CHAPTER XLIX. BRITISH NEEDS AND BRITISH RESOURCES.

LIST OF ILLUSTRATIONS.

CHAPTER I

SANATORIA for consumptives are establishments for the open-air treatment of presumably curable cases of consumption under medical direction, mainly by hygienic and educational measures. They are not intended for patients whose prospects of recovery are small; nor are they intended for non-tuberculous cases, or for those who are merely out of health. Consumptive patients who are seriously ill are better treated in the first instance in a hospital or nursing home, or in their own homes, if these are suitable. Their treatment is quite as much a question of medicine and of nursing as of hygienic measures; and to send them to a sanatorium is to fill up beds which might be more usefully employed in other ways. For the same reason it is inadvisable for a sanatorium to receive people who are merely in need of change of air, as a less expensive health resort, without the special arrangements of a sanatorium, is quite sufficient to restore them to health. Such visitors are only admissible to minister to the comfort or happiness of other patients, or where they have themselves shown a tendency to pulmonary tuberculosis.

The sanatorium is intended to receive the consumptive in any stage which offers a reasonable prospect of recovery, and where there is sufficient power of reaction to render an open-air life possible without mischief. A sanatorium of this kind has therefore a strictly limited though most important function, and is no substitute for the nursing home or hospital, the convalescent home or hydropathic establish-

1

ment, although it resembles each of these in certain respects. Owing to the necessity of being near centres of population most hospitals—even most consumption hospitals—have been erected in situations which are by no means the best for the hygienic treatment of consumptives. Perhaps if they were to be erected at the present day, many would be placed in more suitable neighbourhoods. Some convales-cent homes and hydropathic establishments are quite suit-ably placed for the open-air treatment of phthisis, and might with a little modification be adapted to this purpose. But the class of visitors which they receive, and in some cases the inadequate provision of shelter against wind and weather, would make it difficult to satisfactorily treat con-sumptives there. Many of those who go to a convalescent home do not need the special arrangements of a sanatorium. Hydropathic establishments, too, are intended for a widely different class of visitors, whose dietary is often by no means the best for the average consumptive; and in a different way this is also true of hotels and boarding houses in health resorts. Such establishments are primarily in-tended for the reception of pleasure seekers, whose pre-dominating presence is a serious hindrance to the systematic treatment of phthisical visitors. In a sanatorium the main object is the recovery of health; and for consumptive patients and their advisers this is a sufficiently difficult task, to which everything else must be subordinated.

Sanatoria for consumptives often go through several stages of existence. At first simply health resorts, in which the visitors do whatever seems best in their own eyes without being necessarily under medical advice or guidance; later they become "open sanatoria" for a mixed class of patients, tuberculous and non-tuberculous together, under medical supervision. They are termed "closed sana-toria" when they are entirely under medical direction, and sufficiently complete to provide for all the wants of their inmates; and in many cases a further step is taken by excluding all but phthisical patients. Closed sanatoria possess great advantages in the treatment of consumptives, who are apt with the best intentions to do most unwise actions, and to be led astray by the example of visitors of a different kind, if they wander from the sanatorium into public roads and resorts. The ordinary amusements of a health resort — theatres, concert - rooms, skating - rinks, casinos and even public-houses—may present fatal temptations to the phthisical in an open sanatorium, whereas this cannot so easily happen in a closed one. There are also great advantages in treating a number of consumptives in the same sanatorium, provided they be judiciously chosen ; for the arrangements necessary for two or three will often suffice for a much larger number, and a corresponding saving of time, trouble and expense is thus effected. Nearly all the more recent sanatoria for consumptives erected in Germany are closed sanatoria, exclusively reserved for such patients. Sanatoria for consumptive children require a slightly different organisation from those for adults. In dealing with patients who suffer from external tuberculosis, a much more bracing and exposed situation may often be chosen with advantage. In this country many such cases are treated at our seaside convalescent homes with the best results, whereas these institutions, with present arrangements, are at certain times and seasons for the most part unsuitable for ordinary consumptives. The medical and surgical requirements in the two sets of cases are also usually different.

CHAPTER II.

In his recent book on *The Consumptive and his Treatment*,[1] Dr. Léon Petit says : " We are now-a-days convinced that there is no climate, however favoured, which by itself can cure the consumptive ". I may add that there is no climate which is equally suited to every case of consumption. People in health differ greatly in their powers of reaction, so that the same climate may be bracing to one which is depressingly cold to another. Amongst consumptives there is an even greater difference to be found. At certain stages, and in some cases, a mild equable climate is essential ; whereas for most of the more hopeful cases, a cool bracing climate will be best, although this must be associated with plenty of shelter against wind and weather. It is a mistake to suppose that consumptives generally do best in warm climates. Pulmonary phthisis runs a relatively rapid course in warm climates; and patients who have gained weight in winter often lose it when the warmer weather sets in. In this respect England compares favourably with Italy and the south of France. Consumptive patients from the cooler parts of Germany are said to do badly at the Riviera; and the same thing has been observed of those who go there from the north of France. Some years ago a number of selected cases were sent from the Brompton Hospital to Madeira; but most of them were no better for their change of climate. A similar result is said to have followed in the case of some hospital patients sent from Manchester to Bournemouth. The health resorts

[1] *Le Phtisique et son Traitement Hygiénique*, Paris, 1895, p. 49.

which have been most successful in the treatment of consumptives are almost without exception places which are
cold, or at all events cool, during some part of the twenty-
four hours. This is true of Alpine health resorts, elevated
tablelands in various parts of the world, German hill sanatoria, and many places with marine or semi-marine climates.
Confining our attention to the localities which benefit the
more curable cases, we find they possess certain common
features which we are justified in regarding as essential.
(1) They have a pure air, free from dust and smoke and
the impurities which are inseparable from a dense population; (2) they are fresh and bracing, but well protected
against cold or stormy winds; (3) they have sufficient fine
weather, or sufficient artificial shelter, to render an out-of-
door life possible; and (4) they have a dry, warm, well-
drained soil. Wherever these four conditions are found,
and suitable arrangements can be made, it should be possible
to treat consumptives with success. Neither high altitude,
dry atmosphere, fine weather, equable temperature, nor
abundant sunshine is essential to success, however useful
they may be, or desirable in particular cases. Were high
altitude an essential, we should not hear of recoveries on
the ocean, or in low-lying health resorts. Were fine weather
a *sine quâ non*, the remarkable success of the sanatoria in
the Black Forest and other parts of Germany would never
have been chronicled. Many of these have a moist and
chilly climate during part of the year; but patients do
just as well at such seasons as during finer weather. Indeed
the results are if anything better in winter, and Dr. Dettweiler, Dr. Walther, and other competent observers regard
the weather as of little importance. Moist still air is sooth-
ing to irritable air passages; intense cold, mountain fogs,
snowstorms and the like, need not interfere with treatment if
reasonable precautions are taken. The health resorts which
enjoy the best climates do not usually possess the greatest
amount of shelter from bad weather; but rather those with
a variable and indifferent climate. Visitors have been half-

frozen in Italy and Georgia, and chilled with cold winds at
the Riviera. Of greater importance than the climate is the
use which is made of it; and with the help of artificial
shelters this may be independent even of continued bad
weather. Consumptives have been cured in the most unlikely climates ; and many things point to the conclusion
that it is fresh air and medical supervision rather than a
fine climate which are needed for success.

This statement is not intended to imply that other climatic
characters are of no value ; but rather that they are not of
paramount importance in most consumptive cases. Alpine
climates, for example, are unrivalled in their beneficial influence over certain forms of the disease ; but they are by
no means universally applicable, and other climates can
often with advantage be substituted, for a shorter or longer
time. The value of Alpine health resorts depends partly
on characters which are common to some other stations.
Their power in expanding healthy parts of the lungs can
scarcely be found lower down ; but their tonic influence,
their cool air with a warm sun, their atmospheric dryness
and purity, their absence of dust, their freedom from strong
wind, are possessed in various degrees by places at lower
levels. It is probably easier in Alpine climates to prevent
the growth or continued vitality of the tubercle bacillus outside the body, partly because of the large amount of sunshine, partly owing to the atmospheric dryness and the
winter coat of snow ; but if the sputa are properly disposed
of in every case, this becomes of little importance. It is
most unwise to indiscriminately recommend the Alps to every
consumptive patient. To obtain benefit from such a climate
a certain degree of reactive power is essential : and just as
many people cannot without previous training get a proper
reaction after a cold bath, so it is with an Alpine climate.
Douglas Powell indeed has said : " Notwithstanding the
enthusiasm with which cold mountain climates have been
of late years advocated for the treatment of consumption, it
remains certain that those resorts are not adapted for the

majority of patients, as they come before us, suffering from this disease; and perhaps it might with truth be said that a large number of the very cases that do well aloft do equally well in the plains ".[1] " In order to reap the benefit of high altitudes, patients must be free from pyrexia, and possess sufficient lung surface to carry on adequately the process of respiration in the attenuated atmosphere." [2] Those who are markedly febrile, who have feeble circulation, damaged kidneys, double cavities or extensive lung disease, rapidly advancing lung destruction, laryngeal complications, or irritable nervous systems, do badly at Alpine resorts. Even at Reiboldsgrün, which has not an Alpine climate, being only 700 metres (2300 feet) above the sea-level, it has been found that some patients do not progress so well as at lower levels, and Dr. Wolff-Immermann is in the habit of sending away such as do not show decided improvement within the first fortnight. Most of the existing sanatoria are not placed at high altitudes, as will be seen from the tables given at pp. 84, 129, 149, 206, 267, 273 and 283.

In comparison with Davos and the German hill sanatoria, many health resorts at lower levels are at a disadvantage, owing to the absence of proper provision for the hygienic treatment of consumptives. When this defect has been remedied, it will probably be found that they too, as well as others yet untried, have an important part to play in the battle against tuberculosis. The warm low level health resorts of the Riviera have no longer the reputation they formerly had in this respect; and many think that the climates of some parts of the British Isles would be more useful for our countrymen. " There are many places in this country where, on a dry soil and in a sunny sheltered part, on the southern slope of some upland, most of the conditions can be obtained which are now dearly bought and far sought, and often not obtained, in distant parts of the world." [3]

[1] *Diseases of the Lungs*, 4th ed., 1893, p. 513.
[2] C. Theodore Williams, *Practitioner*, June, 1898, p. 627.
[3] A. Ransome, *The Treatment of Phthisis*, London, 1896, p. 138.

CHAPTER III.

THERE are some very strong reasons for the treatment of consumptives in their own country.

Many patients are quite unfit to take a long journey ; and this is true very often of quite early stages of the disease. So long as there is fever, or fatigue after slight exertion, or a tendency to free perspiration, a long journey must be full of risk. The presence of inflammatory complications is also a contra-indication, or persistent or recent and copious hæmoptysis. To those patients who cannot afford to travel comfortably, a long journey will always be undesirable. In a long land journey the intermediate halting places will often be of a most unsuitable kind. Average hotel accommodation is far from ideal for a consumptive or delicate patient. The larger rooms in even good hotels are frequently left to ventilate themselves ; the bedrooms · may have been previously occupied by visitors suffering from influenza or other infectious complaints. Even hotels which have been specially built for the reception of delicate persons or invalids often leave much to be desired. " It is not too much to say that they are most unsuitable places in which to make an extended sojourn." [1] For dyspeptic consumptives, dietetic treatment is of great importance : but a suitable diet is often difficult to obtain during a journey.

In a sea journey passengers are greatly at the mercy of the weather. The average cabin is too small and ill venti-

---

[1] "The Climate of the Dwelling-house," by G. Vivian Poore, *Journal of Balneology and Climatology*, October, 1897.

lated to be suitable for a consumptive patient, who may in stormy weather be much worse off than at home. There is often great danger of chill in passing from the sheltered to the more exposed parts of the vessel in windy weather. In traversing the tropics the extreme heat is ill borne by delicate passengers. The fare on board ship is often very good for healthy people ; but for many invalids it is by no means suitable ; while the monotony of the voyage and the absence of effective medical supervision lead some to spend an undesirable proportion of their time in drinking and smoking in ill-ventilated saloons. Before we recommend a consumptive patient to take a long voyage, we must take into account his purse, tastes and inclinations, his seafaring capacity, the probability of good weather or the reverse, and his ability and willingness to conform to the necessary hygienic rules.

When the patient reaches his journey's end, it will make a great difference whether he goes to a medically-supervised sanatorium or is left to the tender mercies of chance and his own medical skill. To send a feeble patient to rough it in the colonies is obviously wrong. Equally so would it be to expose him, even in the best of climates, to the temptations of sight-seeing and the injudicious pursuit of pleasure. In either case he would have been far better in his native country. Climate is but one factor in treatment, and accommodation, diet, and above all the use which the patient makes of his time, are at least of equal importance. Without systematic medical supervision, the sojourn in a foreign health resort is nearly always a mistake for a consumptive patient. Dr. Knopf states that in the beautiful climate of Colorado there is an enormous difference in the mortality among consumptives under systematic medical treatment, as compared with those who merely consult a physician when they think it necessary ; and the same is borne out by Dr. Solly's statistics.[1] But even if the patient

[1] *Medical Climatology*, London, 1897, pp. 133-141 ; Knopf, *Les Sanatoria pour la Phtisie Pulmonaire*, Paris, 1895.

can safely travel, and is under medical care, there are still some arguments in favour of treatment in home climates. As a rule (although there are exceptions) it is more expensive to go abroad for treatment; and many patients may consequently have to curtail the time of systematic treatment. National tastes, and even national prejudices, have also to be considered, and the depression which arises when the average patient is separated from his friends and relatives. A more important argument is drawn from the liability to relapse which is shown amongst patients who return to a damp and variable climate from sunnier lands. Many of the foremost physicians abroad are agreed as to the desirability of attempting the cure of consumptive patients in their own country. This is the opinion of Prof. v. Leyden, Prof. v. Ziemssen, Prof. Naunyn, Prof. Senator, Dr. Dettweiler, Dr. Gerhardt, Dr. Fränkel, Dr. Knopf of New York, and others. There was recently a fund started in Germany to establish a sanatorium at Davos for German consumptives of the poorer classes. Germany has no Alpine climates ; while patients in the sanatorium at Davos would probably have been under the supervision of a German doctor ; yet an influential protest was raised on medical grounds against this mode of attacking consumption, and the fund may eventually be devoted to the establishment of other hill sanatoria in Germany itself.[1]

---

[1] *Heilstätten Correspondenz*, Berlin, Dec., 1897 ; Jan. and Feb., 1898.

CHAPTER IV.

SANATORIUM treatment is based on a careful regulation of
each patient's daily life in all its hygienic and medical
details. He is gradually trained to stand a life in the
fresh air in all weathers, while his tendency to chill is
removed by simple hydropathic applications and other
common-sense precautions. The nature and amount of his
daily exercise are regulated according to the weather, and to
his momentary state of health, in an ascending scale, beginning with absolute rest in bed. His food ranges from fever
diet to a rich and varied though digestible dietary. Strict
precautions are taken to prevent all risk of infection, while
the training he receives is useful not only to himself but to
the whole community after his departure from the sanatorium. A resident medical and nursing staff assist him in
carrying out these daily measures, and are immediately
available in case of hæmoptysis, night sweats, catarrhal
and other complications. Not that this constitutes the
whole duty of the medical and nursing staff. They have to
prevent imprudences in some cases, to encourage to per-
severance in others; to strictly enforce all essential rules,
while they allow sufficient personal liberty in less important
matters to prevent the irksomeness of restraint; to suggest
harmless and beneficial forms of recreation, while they dis-
courage all those which are likely to do mischief. Mind
has a great influence over bodily health, and the stimulus
of hope and the encouragement which results from steady-

(11)

progress and sympathetic attention will count for much in curing the patient.

Those who try to live in the open air in a climate like our own will meet with many difficulties, owing to the absence of special shelters and contrivances for warding off rain and wind while admitting fresh air. The ventilation of rooms even in good-class houses is seldom quite satisfactory; and it is quite exceptional to find an adequate provision of verandahs and covered walks for use in wet weather. Such arrangements may not be needed in very favoured climates, but they are essential in an uncertain or a rainy one, especially in dealing with phthisical patients. Consumptives are very apt to over-heat and under-ventilate their rooms, forgetful or ignorant of the dangers which they thus incur. Graduated and increasing exercise is very useful in quiescent phthisis, provided that it is not allowed to excite dyspnœa or greatly accelerate the heart action. But in febrile or dyspeptic patients who are losing weight, all unnecessary exertion is mischievous, and rest in the open air the only permissible method of treatment. The sanatorium and its grounds have therefore to be arranged with a view to both of these contingencies, and paths provided with every variety of gradient as well as sheltered resting places. Dust and organic effluvia are most injurious to phthisical patients. The former excites useless cough and irritates the air passages, while the latter lower constitutional vitality and foster the growth of tubercle bacilli. To banish dust and dirt, special methods of furnishing and decoration, and still more, special methods of cleansing, are necessary, which are scarcely if ever found in ordinary households. Each of these points will be more fully dealt with in later chapters.

CHAPTER V.

A SANATORIUM for consumptives should be placed on a dry soil, in a sheltered situation, with pure fresh bracing air. The best soil is sand or rock, as with a proper fall these are not retentive of moisture, and are consequently soon dry after rain and are warm to the feet. Rising ground should be chosen; with a southerly aspect, and good shelter from hills or woodland against cold or boisterous winds. The question of altitude has already been considered (see chap. ii.). Various elevations according to geographical position have been recommended by Liebermeister, Brehmer, Weber and others. As, however, their reasoning is partly based upon theoretical considerations, it is not necessary to reproduce their figures here. Moreover, as we have already said, many different elevations would probably be needed to suit various degrees of reactive power and lung capacity. The only systematic attempts at sanatorium treatment in England have been at low levels but little above the sea. Nearly all the British health resorts which have a reputation for the treatment of consumptives are on or near the coast. A great advantage of the seaside, also shared by mountain resorts, consists in the recurring breezes which purify the air and increase the bracing effect. Local tuberculoses do exceedingly well on the seashore; but for pulmonary phthisis the immediate and unprotected sea-front is not the best. Dr. Ransome has written in strong terms about the danger of such a situation; and other older authorities—Walshe, Beneke, Fodéré—were equally emphatic. Where the bracing qualities of seaside air are combined with sufficient wind-shelter, consumptives will

(13)

probably do well. Individual peculiarities must, however, not be overlooked. There is a great lack of sheltered yet bracing health resorts in this country with provision for the open-air treatment. Some of those which exist are too relaxing in summer; and others which are more bracing are not provided with sufficient shelter. And if the accommodation for consumptives is inadequate on the coast, it is still more so inland, although many suitable spots could be found for the purpose. Where a high hill is available, it by no means follows that the top is the best place to choose. A slightly lower situation is often both more sheltered and less rainy. Cornet of Reichenhall pointed out[1] that there were often strong winds at the top and at the foot of a mountain; and to a smaller extent this may be true of lesser elevations. Blumenfeld recommends for North Germany the south side of a hill 400 metres (1300 feet) high. The French sanatoria for children at Ormesson and Villiers are on a breezy, somewhat unprotected high plateau, 114 to 121 metres (374 to 397 feet) above the sea-level. The local configuration of the land, as well as the geographical position, have a great influence over climate, and would have to be taken into account.

Purity of the air is of paramount importance. For this reason the neighbourhood of a large town or factory is inadmissible, or the presence of organic refuse near by. One of the drawbacks to the treatment of consumptives in popular health resorts is the increasing contamination of the air from growth of population. As Léon Petit says: " We are now-a-days convinced that there is no climate, however favoured, which alone can cure consumption. The places which are free from tuberculosis are those where the scanty population lives constantly in an atmosphere which has not yet been polluted. They attract invalids, and when they become fashionable they have long since lost the qualities which gave them renown."[2] On the sea coast

[1] Buda-Pesth International Congress of Hygiene and Demography.
[2] *Loc. cit.*, pp. 49-50.

the water and the sea breezes play an important part in counteracting such atmospheric impurity ; but we have already seen that the near neighbourhood of the sea is not the best for consumptives. No sanatorium should be near a high road ; in dry weather the dust is sure to fly up and increase the tendency to useless cough. A large extent of grass land round the sanatorium is an advantage in preventing dust from rising.

High chalk downs have been recommended by some for the treatment of consumptives. They are, however, open to two important objections. In dry weather the short grass of such land is insufficient to prevent the dust from flying up ; while in wet weather the soil remains cold and damp, and the air above it is apt to be foggy. Where hills are covered with a thick layer of sand or gravel they may be suitable for the open-air treatment, provided the area of pervious soil is sufficiently extensive to influence the atmospheric condition. A gravel patch near a marsh or in an ill-drained depression would be unsuitable. The character of the vegetation is a useful guide to the underlying soil. Where pine trees abound the soil is usually sandy and dry ; and these trees are useful in other ways, as they give more permanent shelter against wind and rain, and are active producers of ozone. Some people also believe that their terebinthinate secretions are curative ; although this is not quite so well established. Dr. C. T. Williams has pointed out that where gorse and heather and short springy turf abound, the locality is usually suitable for consumptives, whereas long rank grass should be avoided. It must be remembered that an underlying stratum of clay may render a sandy soil unsuitable for consumptives. Soil and shelter are the two most important points in choosing the site for a country sanatorium.

But if a country place is the most suitable, it must be reasonably accessible to prevent the need for a long journey. The sanatorium should however not be close to a railway.

CHAPTER VI.

THE acreage required for the grounds of a sanatorium will depend somewhat upon the amount of open public land in the neighbourhood. The grounds of the Brehmer Sanatorium at Görbersdorf amount to about 300 acres ; and some of the American sanatoria have very extensive grounds. Land is, however, too expensive in this country for such luxurious proportions; and a much smaller area will suffice if it is of the right kind, and not surrounded with houses or with ploughed fields. Brehmer laid stress on the importance of carefully graduated walking exercise in strengthening the heart of the consumptive; and his results to a large extent justify his contentions. For such a purpose it is necessary to have a reasonable length of level paths, and a number of others at various easy gradients, so arranged that the patient can take walks increasing in length and in difficulty, always returning down-hill to the sanatorium. Since Dettweiler's powerful advocacy of the treatment of consumptives by rest in the open air, the need for systematic exercise out of doors has been somewhat underrated in many quarters ; but there is little doubt that facilities should be provided for both ; and this view is adopted in some of the best German and American sanatoria. At certain stages, absolute rest in the open is the best treatment ; later on, graduated exercise, which should be taken without having to quit the sanatorium for the public roads, as in some states of the weather these may be unfit for consumptives. The grounds should be freely provided with seats and shelters, and covered walks for exercise in rainy weather. In the Brehmer Sanatorium there are seven kilometres (four and a half miles) of walks, and a seat every twenty paces, besides large winter gardens and shelters for wet weather. Trees—preferably evergreen—should be so arranged as to give shelter against wind while they permit the free access of sunshine and air. In these comparatively sunless islands, every ray of sunshine should be utilised to the utmost; for if it be not essential to recovery, it is always grateful to chilly subjects, and the best of nature's disinfectants. In a country sanatorium, stables and other outbuildings are a necessity. They should, however, be placed at a distance from the patients' haunts.

CHAPTER VII.

CONSTRUCTION, DECORATION AND FURNITURE.

THE main facts to be borne in mind are, that the sanatorium is for the open-air treatment of patients, that every part should be extremely well and independently ventilated, readily cleansable without raising dust, and, as regards the patients' quarters, freely open to the sunshine, while sheltered from wind. It is conducive to economy, both in construction and in management, to have a building which is concentrated on a relatively small area. From a purely medical point of view it would be best to have a series of scattered but intercommunicating buildings of few storeys. Existing sanatoria fall into two extreme groups, representing these two ideals, with various intermediate forms. On the one hand we have such a building as Dr. Römpler's at Görbersdorf, or the main building at Reiboldsgrün, or at Leysin ; on the other we have the Cottage Sanatoria at Nordrach in the Black Forest, the Adirondack and Loomis Sanitaria in America. One of the chief difficulties in designing a sanatorium for consumptives consists in the advisability of providing a sunny aspect for every patient's bedroom. In an ordinary hotel or boarding-house the rooms may face every point of the compass : they often block the ends of corridors ; while space may be further economised by arranging many rooms round the sides of a hollow square. But such an arrangement is inadmissible for a sanatorium. The plan adopted at Falkenstein is that of a central five-storey building with diverging wings, protecting a terrace round which are placed the deep verandahs for fresh-air treatment. Diverging at a still more open

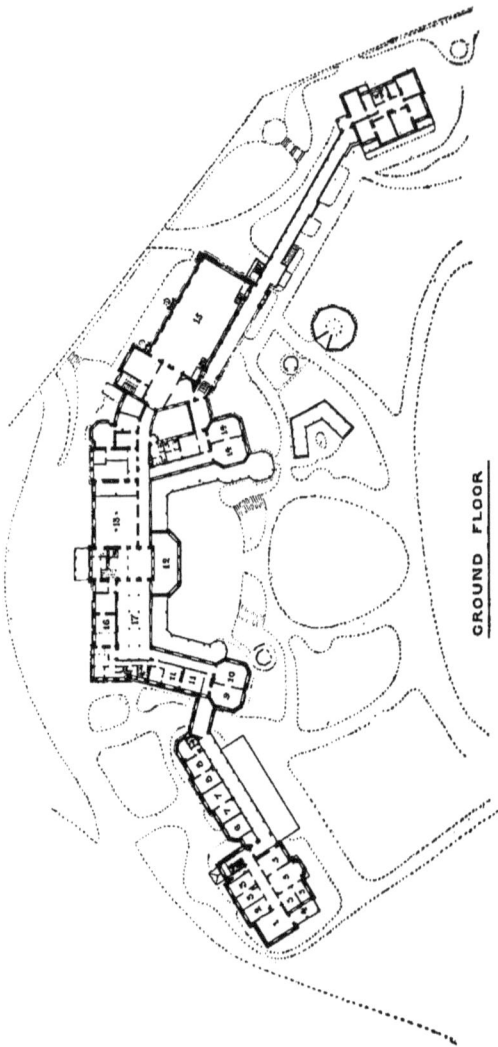

GROUND FLOOR

FIG. 1.—THE FALKENSTEIN SANATORIUM.—GROUND PLAN.

1. Board Room.
2. Visitor's Room.
3. Rooms for the Staff.
4. Verandah.
5. Consulting and Waiting Rooms
   and Laboratory.

6. Lavatory.
7. Mortuary.
8. Gardener's Quarters.
9. Library.
10. Billiard Room.
11. Visitor's Rooms.

12. Winter Garden.
13. Reading Room.
14. Conversation and Music Rooms.
15. Dining Saloon.
16. Office.
17. Winter Garden.

[Face page 19.

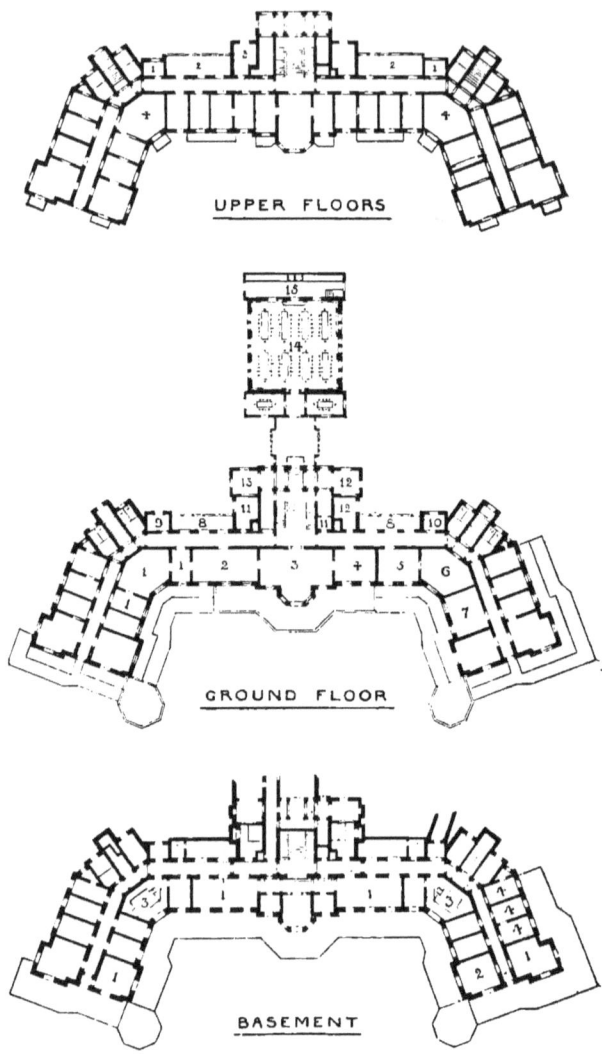

UPPER FLOORS

GROUND FLOOR

BASEMENT

FIG. 2.—THE HOHENHONNEF SANATORIUM. [Face page 19.

The Hohenhonnef Sanatorium.

Basement :—

1. Cloakrooms.
2. Inhalation Room.
3. Heating Apparatus.
4. Douche Rooms.

Ground Floor :—

1. Consulting and Waiting Rooms.
2. Winter Garden.
3. Reception Room.
4. Reading Room.
5. Ladies' Room.
6. Music Room.
7. Billiard Room.
8. Cloakrooms.
9. Servants' Room.
10. Nurses' Room.
11. Hairdressing Room.
12. Office.
13. Post Office.
14. Dining Saloon.
15. Serving Room.

Upper Floors :—

1. Nurses' Room.
2. Balcony.
3. Small Kitchen.
4. Rooms for Convalescents.

angle are covered corridors (one open to the weather), which
lead to later-built annexes.   Immediately behind one of
these corridors is the separate large dining saloon,
which is over the kitchens (see fig. 1).   The same plan
has been adopted at Hohenhonnef, with the exception of
the lateral corridors and annexes, and the position of the
dining saloon (see fig. 2).   In such an arrangement
nearly all the parts used by patients are on the south
side.   At Hohenhonnef, for instance, the verandah is placed
in front of the basement, which is occupied by cloak-rooms,
store-rooms, and the like.   On the floor above are the reception-rooms ; above this are placed the chief of the patients'
rooms on three floors, a few more being situated on other
aspects in the wings.   Such tall buildings make a lift almost
a necessity, although at Falkenstein this has not yet been
provided.   Another possible drawback is that each floor is
apt to ventilate into the one above, unless special precautions are taken.

In the cottage sanatoria, on the other hand, most of the
buildings consist of just two (or at most three) storeys—a
ground floor, which may be occupied by reception-rooms
or (if raised and damp-proof) by some of the patients' bedrooms.   The kitchen and its adjuncts would form a separate
building, and there would also often be a recreation pavilion.
This is an ideal plan where land and material are cheap, and
there are plenty of suitable spots for a number of buildings,
and the weather is fine during most of the year.   Where
these conditions are not present some compromise between
the two systems must be adopted, or at all events the buildings must be connected by covered ways.   The cottage
system is the more convenient for isolating such patients as
may catch an intercurrent infectious disease ; and also for
grouping patients according to the form and stage of disease
where there is much diversity in this respect.   It is however
more expensive and more difficult to supervise.

Verandahs and covered balconies are an essential part
of every sanatorium for consumptives.   These should be
so arranged as not to darken the patients' rooms.   Glass

roofs will often be a help in this respect; or the verandahs may be placed at a lower level or to one side, as at
St. Blasien and at Ruppertshain (see fig. 29, p. 254). There
should if possible be at least one verandah or summer-house
with a northerly aspect for use during hot weather, which
may otherwise be exceedingly debilitating to phthisical
patients. To properly shelter the patients, the verandahs
should be twelve feet wide or more. In rainy climates they
should be provided with movable glass screens, as at
Hohenhonnef and the Adirondack Cottage Sanitarium.
It would be a great convenience if the verandah floors
could be warmed in winter. They should in every
case be sloped, to throw off moisture.

· The kitchens should be placed at a distance from the
haunts of the patients, to prevent loss of appetite from the
smell of cooking. Where this is not possible, they must be
cut off by ventilating lobbies and corridors. Cloak-rooms
and lavatories should be readily accessible to patients on
their way from the grounds to the dining-saloon to prevent
unnecessary fatigue, and the cloak-rooms should have hot
pipes or drying cupboards for damp clothes. In other
respects the buildings must ·be constructed according to the
accepted rules of hygiene. Dry foundations are of course
essential, as well as adequate provision of suitable closets,
bath-rooms and the like. Most of the German sanatoria
contain a douche-room, which may be placed in a lower
floor on the northern side. At Nordrach there is a douche
apparatus in each of the bedrooms. In some sanatoria
this method of treatment is not even used. The provision
of baths in foreign sanatoria is sometimes inadequate,
although there are notable exceptions. Waste pipes from
lavatories and bath-rooms become almost as foul as soil
pipes from water-closets. They should therefore be placed
in built-out pavilions, and be as short as possible. It
would be an improvement to provide for their cleansing by
mechanical means, as this is not properly accomplished by
even a powerful flush of water.

The ventilation of rooms should be as perfect as possible. It is quite impossible to rival the open air in its purity, but as near an approach should be made as is practicable. The ordinary allowance of 3000 cubic feet of air per head per hour is often regarded as a "counsel of perfection"; but for open-air treatment this is not sufficient. With open windows and open chimneys or ventilating shafts, more than this can be provided even in winter as soon as patients have learned to stand a very moderate degree of cold. Dr. Ransome states that at the Manchester Consumption Hospital as much as 18,000 cubic feet per head per hour was often provided:[1] and with the help of special arrangements there should be no difficulty in improving upon the ordinary standards of indoor ventilation, high as these are above those observed in most of our private houses and hotels. In many continental sanatoria the windows are kept open in all weathers and seasons for the whole twenty-four hours. Where the climate is variable, special shutters are often provided (as at Ruppertshain Sanatorium) which let in air while they exclude rain. Small glass screens might be arranged for the same purpose. Some sanatoria have double windows. This is unnecessary and may be mischievous, as they can be of no use unless the windows are shut. Even in Alpine sanatoria it is doubtful whether they are necessary. All windows and shutters should be so constructed as not to clatter when the wind blows. I have known it to happen that with ordinary fittings no sleep was obtainable until the windows were shut. Abundant light should be admitted into all the rooms, and all those used by the patients should admit the sun's rays during part of the day. The main front of the building should therefore be south, south south-east, or south south-west. People are apt to forget that the eastern side of a house receives more sun than the west, especially in hilly districts. If there is adequate shelter against east winds,

[1] Weber-Parkes Prize Essay, London, 1898, p. 73.

the east side may quite properly be used for some of the bedrooms. In summer time it is unpleasant to have the hot noonday sun streaming into the room. A direct southerly aspect is therefore not so good in summer for any patient who is confined to his bed. This difficulty may, however, be overcome by wheeling the bed on to a balcony or verandah with a different aspect; and in the same way a cold aspect may be rendered suitable for a patient in winter. In such cases the bedroom would be used solely during the night. In the main building at Nordrach, the noonday sun falls aslant along the roof in summer, but shines into the bedrooms in winter.

According to Trelat (Buda-Pesth International Congress) the light on a dull day indoors should be at least equal to that of ten-metre candles. At least half of one side of each room should consist of window space. Many people cannot sleep unless the bedroom be darkened. This should be effected by the outside blinds rather than by thick curtains, which are apt to hold the dust.

Artificial lighting should be by electricity, as other methods add to atmospheric impurity. At Hohenhonnef even the open-air verandahs and shelters in the woods are lit by electricity. More primitive methods are however employed in some otherwise good sanatoria.

The cubic space is of less importance than the size and position of the ventilating openings, although, of course, within certain limits the larger the rooms are made the better. The large dormitory at the Villiers Sanatorium for children (p. 141) contains over 4200 cubic feet per head. At Hohenhonnef none of the patients' rooms are smaller than 2470 cubic feet. At Nordrach, where very good results are obtained, they average 1500 cubic feet. At Albertsberg and Oderberg (both for the artisan class) from 1200 to 1400 has been allowed. Patients in sanatoria spend most of their time out of doors, so that with adequate ventilation 1500 cubic feet should suffice, with a minimum of 120 to 150 square feet of area in the bedrooms. Heating in the continental sanatoria is usually by low-pressure steam, which is probably the cheapest method and one of the most easily regulated. Charring of dust particles by contact with the heated pipes may be prevented by reasonable care. In some cases hot-water pipes, closed stoves, or, in America, open fires are employed. On the continent chimneys are by no means always present in all the bedrooms, although they are usually provided in the newer sanatoria for paying patients. The absence of a chimney or a corresponding air-shaft is of course a great drawback. Open fires are most admirable ventilating contrivances. They are, however, relatively expensive; they demand more attention from servants, and there is a little risk of dust blowing about in windy weather. Closed stoves must be very carefully constructed to be admissible in a sanatorium. In the Adirondack Cottage Sanitarium the patients' rooms open out of a common sitting-room, being separated by an incomplete wall seven feet high, the part next the ceiling being open, and the doors having a ventilating space below. This considerably increases the air-space, and makes it easier to ventilate in doubtful weather without draught. Where the incoming air is warmed, care must be taken not to overheat it. The temperature in some of our consumption hospitals is often too high for consumptives, although suitable for ordinary bronchitics. This is one of several reasons for separating the different kinds of patients in chest hospitals. At the Loomis Sanitarium the winter temperature in the bedrooms at night is kept under 40° F. It is highly desirable that each room should be capable of being ventilated from more than one quarter, according to the direction of the wind. It is essential that they should be flushable with a direct current of fresh air. This is the chief reason why the bedrooms of a good sanatorium are arranged in a single row with a corridor behind. The windows of the corridor should be opposite the door of each bedroom. In summer time ventilation may be greatly improved by removing the doors. With the bed in a sheltered position both door and window may often be safely left open. For the sake of privacy, and to check the too rapid passage of air, skeleton doors with permeable centres may be substituted for the solid ones. This has been done at my suggestion by Miss Jane Walker, M.D., in a small sanatorium in Norfolk under her supervision, "greenhouse shading" of double thickness being used. The material should of course be removable for periodical cleansing. A similar contrivance may with advantage be adopted where the windows do not reach the floor, the permeable screen replacing the wall below the window frame. The inlet may in this case be made smaller outside than in, on the principle of Ellison's perforated bricks. With a wind screen an amount of ventilation will often be possible which would otherwise not be tolerated. At Nordrach draughts are entirely disregarded. It is worth remembering that a draught often disappears when the ventilating inlet is enlarged.

Every part of the sanatorium should be so constructed as to facilitate cleansing. Corners should be rounded, and unnecessary ledges avoided. The floors may be of polished wood, or painted, or covered with linoleum. Cracks and joins should be as far as possible avoided. By waxing, the cracks may be often filled in. A paraffin composition is also used for the same purpose. In bath-rooms, kitchens and corridors, cement, artificial stone, or tiles may be used; but these are less suitable for parts which cannot be well warmed. At Sülzhayn many floors are made of "torgament," a kind of cement which unites with wood or iron, and has incorporated with it wood shavings, so that it is both cleanly and warm to the feet. Walls may be lime- or colour-washed; or panelled with wood and varnished, oiled, or painted; or covered with lincrusta or other washable materials. If wooden panelling be used, cracks must be carefully avoided. Cement walls, as in most of our hospitals, would be quite suitable. In some continental sanatoria wall paper is used, but this is not advisable unless it be washable. In the Basel Sanatorium at Davos an ex-cellent linen-backed washable paper is used. Silicate paint and other impermeable coatings might be useful in covering inner walls. The ceilings in most continental sanatoria are white-washed or colour-washed, and free from mouldings. At the Maria Sanatorium at Halila in Finland they are panelled with wood; and the same has been done at Nordrach. In planning a sanatorium, care should be taken to place all noisy portions at a distance from the patients' rooms and resorts. Even apart from this reason, it would not be advisable to place the kitchen department, scullery and servants' quarters, next the open-air verandahs or bedrooms. In the "Home" at Denver, the walls are made extra thick and solid in order to prevent one patient from being disturbed by another's coughing. But in a sanatorium it is chiefly the early cases that would be received, and by suitable training, together with simple harmless remedies, all unnecessary coughing can in such patients be usually prevented. Moreover, where this cannot be done, as the windows have to be kept open, the sounds would still travel from room to room. A large building with long bare corridors is very difficult to keep quiet. Many of our otherwise admirable hospitals fail lamentably in this respect. Where thick curtains are permissible, these will help to prevent the echoing sounds from travelling: but in a sanatorium for consumptives this is not advisable, and in any case it adds to the work of management. At Hohenhonnef the bedrooms are provided with double doors to shut out the noise. A well-planned smaller building has a great advantage in this respect.

But few common rooms need be provided in a sanatorium, as it is advisable to encourage patients to stay out of doors. Recreation pavilions and the like should be open-air structures: and even the dining saloon may be usually kept open along one side. In an English sanatorium it would be idle to slavishly copy any of the foreign institutions. Climate and national customs and prejudices have to be considered ; and what is suitable in sunny Colorado or snowy Switzerland would be out of place in this country. The Alpine sanatoria have to be made on a more concentrated plan than those in the Lowlands, in order to economise warmth; and verandahs, which leave rooms sufficiently lighted abroad, would sadly darken those in our less sunny country. The class of patients for whom the sanatorium is designed will also somewhat modify the plan of construction. Every sanatorium should be suitably provided with drug-room and laboratory for bacteriological and other work. Most of the more recently constructed ones have a room with apparatus for radiographic work. There should be means for isolating and nursing the graver cases, and a room available for use as a mortuary in case of need, even though only early cases are admitted.

The furniture of the sanatorium should be light and free from unnecessary ornamentation or dead space: the heavier articles readily movable for cleaning purposes and capable of being cleaned all over, as well as the floor on which they stand. If stuffed furniture is used, the stuffings should be removable, or else kept covered with washable covers. Leather-covered lounge chairs would, however, be unobjectionable ; but saddlebag covering is unsuitable. Bentwood cane chairs are good. For the open verandahs and summer houses the most convenient couches are adjustable cane ones, convertible into chairs by removal of the foot piece. There is no reason why they should not be provided with comfortable cushions, provided these have washable covers. Tuberculosis is not like scarlatina in its infectious properties. It has been suggested to me that the furniture should be capable of being "stoved" *en masse;* but this would necessitate a very expensive method of construction, or a great lack of comfort, and appears to me quite unnecessary. Carpets, mats, curtains and other hangings should be only sparingly used. The carpets should be in movable strips, with a free margin of floor around the room. Hairy mats and long-pile carpets are not advisable. Skins are also best avoided, as leather is not easy to disinfect by heat without damage. Screens and hangings need not, however, be entirely banished, as they are useful in preventing draught and noise, and if of washable materials are otherwise unobjectionable.

Plain iron bedsteads are the best kind for a sanatorium. In some of the continental sanatoria horsehair mattresses are used with covers to button, so that the contents can be taken out and separately sterilised by heat. At Sülzhayn (an institution for those with small incomes) the horsehair is mixed with one-third sissal fibre, which is springy and clean, and considerably cheaper than horsehair. The pedestals are often made of open japanned ironwork with a glass top, which is better than the usual wooden kind. As it is undesirable to have bags and boxes in the bed-rooms, cupboards or wardrobes of some kind are necessary. In some sanatoria these are placed outside in the corridors. Cupboards encroach less upon the cubic space of the room than wardrobes. Their floors should not be sunk below the floor of the room, and the interior should be made as carefully as that of the room itself. The chief objection to their presence is that they may escape the daily cleansing which is so important for every part of the room. Boots and shoes (beyond those actually in use) are best kept in a special well-ventilated boot-room. Outdoor clothing can be kept in one of the cloak-rooms. For washstands, a choice may be made from several of the ordinary types of marble-topped or japanned iron ones. It is customary in many of the sanatoria for the poorer classes for all per-sonal ablutions to be done in lavatories or bath-rooms. At the projected sanatorium at Mont Bonmorin (p. 136), no decorations or pictures are to be permitted beyond those painted on walls and ceilings. Elaborate picture frames and intricate ornaments are, of course, unsuitable for a sanatorium ; but the simpler kinds may quite safely be admitted under good management, although they add to the work of the establishment.

CHAPTER VIII.

ALTHOUGH tuberculosis belongs to the infectious diseases, it is probably the easiest to prevent from spreading. Simple cleanliness alone will go a long way to prevent its extension to other people ; so that many were doubtful whether it was really infectious, until Koch and others proved it to be so. The consumptive in all but the earliest stages is, however, a source of danger to those around him, unless he adopts certain simple precautions. The danger lies not in his breath, but in what he coughs up. It has been calculated by Hiller that there may be as many as 300,000,000 tubercle bacilli in each expectorated morsel of 3 c.cm. Were these sputa to be allowed to dry and mingle with the dust, they would constitute a serious danger for every susceptible person who happened to inhale them. Nor does the danger stop here ; for Ransome has shown that the bacilli are capable of growing in ordinary temperatures on damp wall-paper, especially in presence of organic effluvia from the breath or the soil. Fortunately, sunlight and fresh air are most efficient disinfectants ; and if the sputa are received into suitable receptacles and destroyed before they dry, there can be no possible danger (see also p. 65). In most continental sanatoria Dettweiler's portable spitting flasks are used by the patients ; various kinds of spittoons are also employed indoors, and at Falkenstein are placed about the grounds. This latter plan is however likely to defeat its own ends. Dettweiler's flask has

(28)

done most important services, but is too complicated to be perfect, and is rather expensive. Spitting flasks with one opening should have it large enough to admit of proper cleansing by mechanical means as well as by boiling. They should be free from corners and angles and cracks, preferably of one material throughout the interior, so that no junctions exist in which dirt might accumulate. They may with advantage be opaque or semi-opaque. They should be readily opened, but not liable to leak; and should be of a convenient shape and size for the pocket. Spitcups for the bedside or shelf should not be readily upset. They might with advantage be made with hingeless automatically closing lids. Spittoons for the floor should never be used in a sanatorium. In some continental establishments elegant vases of coloured glass are placed instead of spitcups on shelves in the reception rooms, and on the pedestals in the bedrooms. Another method, adopted in some American and German sanatoria, as well as in England, is to use cuspidores with paper linings, which can afterwards be burnt.

Handkerchiefs of the ordinary kind are banished from most sanatoria; in some, however (*e.g.*, Sharon Sanitarium), patients are allowed to use them, keeping them in india-rubber pouches, and frequently changing them. Detachable linen linings to the pockets would answer the same purpose as these pouches. It is often difficult to induce patients to use spitting flasks; and in any case, it is often necessary to have something for wiping the mouth. For such purposes Japanese paper-handkerchiefs are convenient, as they are cheap and tough, and can be readily burnt. They may also be used instead of napkins at dinner-time. At Falkenstein each patient has his own linen napkin at table, in a linen pouch marked with his number; but the other plan is preferable. Bedridden patients should not be allowed to keep handkerchiefs under their pillows or under the sheets, but in some cleansable receptacle outside the bed. Whatever method is adopted, it is most essential that the sputa should be exclusively received into proper receptacles and kept moist until they are destroyed. In all the foreign sanatoria there is a strict rule against promiscuous spitting. In Germany this is enforced on pain of expulsion. At Heiligenschwendi, in Switzerland, a small fine is substituted; but this appears to me to be mistaken policy. To keep the sputa moist plain water may be used, or carbolic solution (5 per cent.), perchloride of mercury (1 in 1000), or lysol. At Hohenhonnef a solution of lysol and soft soap is used, as this makes the sputa less repulsive to look at. The most efficient methods of sterilisation are burning or boiling, although in some places disinfectants are depended upon. At Oderberg and elsewhere the sputa are mixed with peat mould or sawdust and burnt, the receptacle being boiled. To disinfect sputa by boiling, this must continue not less than five minutes.[1] In some establishments (Oderberg, Sülzhayn) there are special rooms near the W.C.s for cleansing the spitcups. At Nordrach no special disinfection is attempted, the sputa being merely washed down the W.C. This does not seem to be a good plan, although it is fair to remember that the water supply at this sanatorium is unusually abundant and at high pressure.

Linen used by a patient should be disinfected by steam or by boiling. Some sanatoria have steam laundries of their own. It has been suggested to me by Sir J. Crichton Browne that the drying-rooms might usefully be made with blue glass roofs, through which the chemical rays of the sun would penetrate. Covers to furniture and similar articles should undergo regular disinfection. The rooms and corridors should be daily cleansed by means of damp cloths, which could afterwards be boiled. The ordinary methods of broom and duster are not permissible, as they merely distribute the dust through the air. It is usual on the departure of a patient to disinfect the room he has occupied. The walls, if covered with ordinary paper, may be rubbed

---

[1] Schill and Fischer, *Mitth. a. d. K. Gesundheitsamt*, Bd. ii., 1884, p. 133.

down with bread, which is afterwards burnt. Dr. Bowditch
at Sharon Sanitarium has the paper stripped and replaced.
Colour-washed walls may be rewashed. It is best however
to have the wall surface of washable materials which can
be cleansed with soap and water. In France and in some
parts of Germany it is usual to employ a spray (such as
Gereste and Herscher's) with solution of corrosive sublimate (1 in 1000). To be efficacious this must be freely
applied, as organic matter reduces the mercurial salt.
Another useful disinfectant which may sometimes be used
is chlorinated lime solution. In some sanatoria formalin
vapours are employed. In this case the surfaces to be disinfected must be previously ·moistened. Soap and water
have been found quite sufficient for the disinfection of rooms
in our own consumption hospitals and in some of the
foreign sanatoria. Bedding of all kinds should be steam
disinfected after a patient leaves the building. It would
hardly be necessary to disinfect bedroom slops. To scald
out the utensils with a little hot disinfecting solution should
suffice. No precautions are usually taken concerning the
sewage in sanatoria of this kind. Every patient is instructed for his own sake not to swallow the sputa : and
grave cases are not usually admitted. There is no certain
evidence as to the fate of the tubercle bacilli in sewage.
So long as they remain moist there would be no danger,
and their vitality would probably be exhausted before they
could do any harm. Where the dry-earth system is adopted
the product should be freely exposed to sunshine for several
days before being spread abroad. Strong disinfectants could,
it is true, be added to the excreta ; but it is very doubtful
if this be necessary, while the method is open to some practical objections.

The reasons in favour of one or other method of sewage
disposal are not altered by the fact of a sanatorium being
inhabited by consumptives. In some places the water-
carriage system would be best : in others dry systems are
preferable In country places the water-carriage system

would require a larger area of land ; but the waste water from kitchens, etc., would in any case be applied to the land in some way.

That a pure and abundant water supply is necessary at a sanatorium is obvious. For the douche a pressure of one atmosphere is advisable, and a temperature of 10° to 14° C. at all seasons; but in many sanatoria the douche is not employed, being replaced by cold ablutions and cold frictions.

CHAPTER IX.

THE FRESH-AIR TREATMENT.

An open-air life is the keynote of sanatorium treatment. The patient, instead of being kept in a carefully warmed room ventilated from other parts of the house, according to the popular notions of old, lives in the open air from morning till night, at all seasons and in all weathers. Lack of fresh air is the greatest predisposing cause of consumption; fresh air is the most potent means of restoring him to health.

Now this prescription is very easy to carry out in dry climates such as those of Egypt, the Alpine health resorts, South Africa or Colorado; but requires special arrangements and special precautions in a damp and rainy climate. The credit of showing how this may be accomplished belongs mainly to Brehmer, Dettweiler and their followers. The open-air method may perfectly well be carried out in any climate which is healthy for those who are not consumptive. As Léon Petit observes: "Here the climate may help the cure, . . . there it may hinder it; but it only exerts a secondary influence on the treatment".[1] Moreover, just as the pleasantest climates are not always those which are best for healthy people,[2] so it may be that the most pleasant climates for an out-of-door life—where the air

---

[1] *Loc. cit.*, p. 49.

[2] Hermann Weber and Michael G. Foster, article in *Allbutt's Syst. of Med.*, on "Climate in the Treatment of Disease".

is warm and dry, and little rain falls—are not the best for
those consumptives who have later on to return to a less
favoured place. It is bracing climates rather than warm
and equable ones which have the greatest influence in
restoring the consumptive to health, in all but exceptional
cases.

For the open-air treatment, a fourfold shelter should be
provided against wind, excessive cold, extreme sun heat
and rain. Wind raises dust, increases cough and dyspnœa
in consumptives, and intensifies the chilling effects of cold.
The foreign sanatoria with few exceptions have both
natural and artificial shelter against wind. Cold within
certain limits is useful to the consumptive ; but it should
be a windless cold, and suited to the individual power of
reaction. As damp intensifies the climatic effect of both
heat and cold, the chilly consumptive will be able to withstand a lower temperature in a dry than in a humid climate.
Protection against rain and snow will seem to most people
an obvious necessity, although at Nordrach rain is often
disregarded. It is not enough to provide resting places in
the sanatorium which are protected against rain. At certain
stages exercise is imperative, and sheltered paths and open
covered corridors are needed for exercise in rainy weather.
At Falkenstein there is such a corridor 200 feet long.

In places where the sun's rays are very powerful, as
at Canigou in the Pyrenees, direct exposure to the sun is
found to increase the tendency to fever. Even at Hohenhonnef on the Rhine, which is not far south, a large verandah
has been provided which in hot weather can be artificially
cooled by a stream of water. Protection against wind and
weather is afforded in most sanatoria by large verandahs,
which may be fitted with movable glass screens, as at the
Adirondack Cottage Sanitarium. In our own climate it
would be useful to have a verandah with a hollow floor which
could be warmed, as cold and damp can be much more easily
borne if the feet are kept warm. Other simpler ways are the
provision of hot bottles and warm clothing. Recumbency

also helps a chilly patient to keep warm, the blood circulating with less cardiac effort in this position. According to Dr. Weicker and Dr. Jacoby the recumbent position also favours the flow of blood to the apices of the lungs. In most foreign sanatoria summer-houses or sun-boxes are also provided ; at Falkenstein, some of these can be rotated according to the direction of the wind. Dr. Burton-Fanning, in his experimental sanatorium near Cromer, has modified the well-known shelters of our sea-side resorts by providing them with reversible glass screens (see *Practitioner*, June, 1898, and *Lancet*, March, 1898).

Patients who begin their open-air treatment in wintry weather usually need a short period of acclimatisation, during which the time and extent of exposure are gradually increased. Evening air is not found to be injurious to such patients as are trained to an out-of-door life ; and they are provided with artificial light so as to be able to sit out after dark and read their books and papers, even where the climate is moderately humid. Rest out of doors in the verandahs is practised in all but the coldest, wildest weather. Thus at Falkenstein, patients have been able to stay out of doors even during thick fog, during snowstorms, and when the temperature was 10° or 12° C. below freezing. As many as 40 per cent. were able to stay out seven hours or more ; and 6 per cent. could stay out ten or eleven hours a day.[1] Blumenfeld, at the same sanatorium, made an elaborate investigation of the number of patients confined to their bedrooms under various meteorological conditions, and could find no noticeable difference, except during the pre-valence of north-east winds.[2] This is somewhat remarkable when it is remembered how readily an untrained consumptive becomes chilled. It is no doubt in part attributable

[1] P. Dettweiler, *Die Behandlung der Lungenschwindsucht in geschlossenen Heilanstatten.* Second edition. Berlin, 1884.

[2] F. Blumenfeld, *Ueber den Einfluss meteorologischer Vorgänge auf den Verlauf der bacillären Lungenschwindsucht.*

to the still air usually prevailing at Falkenstein, but largely also to the careful individual training and attention to details.

In many German sanatoria there is a systematic application of a simple kind of hydrotherapy, which trains the skin to withstand changes of temperature. At first, patients are rubbed with a dry sheet; then with spirit; then with spirit and water; in the next stage a wet sheet is used; and in those who are sufficiently prepared, cold sponging and douching. A divided cold douche acts as a powerful stimulant to the skin, and through it to the rest of the body, increasing the tissue changes, improving circulation and digestion, as well as the tone of the skin itself. Its stimulating properties depend within certain limits on its coldness, and the force and sub-division of the stream. It was formerly much used in sanatoria for consumptives, but owing to some bad results in a few cases, has been less and less employed, being only occasionally used in some sanatoria and not at all in others. If used at all, it should only be applied for a very few seconds. The other simpler applications are not open to the same dangers, and, although less powerful, are probably sufficient in most cases. There is a strong objection at Falkenstein to the use of hot baths, which are only occasionally permitted, and never very hot or for more than a few minutes. The contention is that such baths relax the skin and debilitate the patient. But the experience of very hot baths in Japan and elsewhere would seem to prove that they may have stimulating effects. The matter is one which should be decided by individual experience rather than by *a priori* reasoning. At Nordrach, afebrile cases are allowed to freely douche themselves with water of any comfortable temperature; but are enjoined not to actively dry themselves.

The clothing of consumptive patients should be no thicker than is necessary to keep them warm. More than this relaxes the skin, and increases the tendency both to

profuse perspiration and to catching cold. The clothing
should be evenly distributed, and not too heavy. It should
not hamper the movements of the chest. Woollen gar-
ments are usually recommended next the skin, mainly
because, by their hygrometric qualities, they keep the skin
dry.   Dr. Walther, however, prefers a less irritating material for underclothing; and there is something to be said
for his contention.

CHAPTER X.

REST AND EXERCISE.

IF the fresh-air treatment merely consisted in letting every patient rest in the open air, or follow his own inclinations as to the amount of exercise, it would be an exceedingly simple matter, scarcely requiring medical supervision. But the essence of Brehmer's and Dettweiler's methods is the elimination of haphazard treatment and the prescription of absolute repose or of various degrees of exercise according to definite medical indications. Patients who are febrile must be kept at rest ; if persistently febrile or with high temperatures at night, absolute rest in bed is needed, windows being kept open, or the bed wheeled on to the balcony according to weather and season and other indications. Where there is only moderate pyrexia during part of the twenty-four hours, they may be allowed to come on to the verandah and rest there on reclining chairs. In this way the fever is usually reduced, and the appetite and other symptoms improve. Another class of patients who are best kept at rest are the decidedly dyspeptic, and those who are rapidly losing weight. If the nutritional income is less than the expenditure, the latter must be as far as possible reduced. The same is true of those who are very anæmic or very feeble.

In all other cases exercise is needful, according to an ascending scale. We may begin with passive motion, or very gentle massage, followed by resisted movements in the recumbent position. After this, very gentle walking exercise may be tried, at first restricted to a few yards at a time on level ground. As the invalid gets stronger the length of his walk is gradually increased ; he then attempts a gentle

uphill walk ; and in time he is able to take long walks up even
steep hills with advantage. He is however never allowed
to walk quickly ; about two miles an hour is a very good
regulation pace for most. He must never put himself out
of breath by his exertion, and should stop directly he begins
to feel fatigued. Prolonged fatigue, profuse perspiration from
the walk, a decidedly raised pulse rate or respiratory rate
show that the exercise has been too much. If patients come
in chilled from their walk, brisk friction with a warm towel,
or a little digestible warm food, will often set them right.
With few exceptions patients should rest for half an hour
or more before every important meal and for an hour after.
The most active exercise is best taken in the morning.

The same rules must guide us as regards the occupation of
the patients. Sedentary occupations which contract the chest
are usually forbidden : in German sanatoria knitting and
sewing are not allowed, for this and other reasons. Indoor
occupations must be restricted as much as possible ; but
with a little management many things usually done indoors
may be carried out in the open air. Thus periodical concerts, recitations, tableaux vivants, and amateur theatricals
may take place out of doors during a great part of the year.
At some sanatoria there are fortnightly social evenings ; at
others, various societies for reading, music, chess, and other
quiet games, or for botany, photography, and other scientific or artistic pursuits, are formed amongst the patients.
Nearly all the continental sanatoria have well-stocked
libraries ; and as verandahs and summer-houses are lighted
up after dark patients can read out of doors until bed time.
Fairly robust patients may be allowed to indulge in out-
door games. Croquet is often permitted ; tennis is looked
on with less favour, as it is seldom advisable, until convales-
cence is far advanced, to exert the arms too freely, owing
to their close connection with the chest. Damaged parts of
the lung should as far as possible be kept quiet, although it
is useful to freely expand the sound parts when the disease
is quiescent. Sledging is a recognised amusement at the

Alpine sanatoria; but this, and still more skating, are somewhat risky for most patients. Gentle cycling on level ground involves far less exertion, and may more often be permitted. Quiet drives are also often permissible. Walking however is the most universally applicable form of exercise, as soon as there is no more fever.

When the disease is quiescent many authorities lay stress on the value of respiratory exercises, which consist mainly in methodical deep inspirations together with simple arm exercises calculated to expand the chest. Waldenburg's pneumatic cabinet is occasionally employed for similar purposes. At the Colorado Sanitarium systematic gymnastics are a recognised method of treatment for convalescent consumptives. Without careful supervision, however, these methods might easily do more harm than good.

In the earlier stages of treatment rest is far more important than exercise ; and this should include mental as well as physical repose. Mental activity necessarily involves expenditure of energy which is needed for the repair of damaged lung tissue and the restoration of constitutional strength. Those who are familiar with the Weir-Mitchell method of treatment will know the importance of economising strength in feeble patients by the avoidance of emotion and of active exertion of all kinds. For this reason it is better to abstain from all business, especially of a worrying nature. At Reiboldsgrün exciting books are marked with a star and forbidden to febrile patients. At Nordrach mental repose is persistently aimed at, the visits of friends and relatives being discouraged. As soon as fever has completely disappeared the patients take walking exercise regardless of rainy weather : but are only allowed to go in small parties of not more than three or four. The reputation of Dr. Walther stands so high that one cannot regard such rules as of no account ; but they seem to be somewhat unnecessarily stringent. Still it is impossible to forget the effect of " visiting day " in raising temperatures amongst the inmates of a general hospital; and such precautions are based upon perfectly sound principles.

CHAPTER XI.

THE food of consumptive patients must conform to the ordinary rules of dietetics. It must be of a mixed character, containing representatives of the different classes of food stuffs; it must be digestible, appetising and varied; it must be sufficient to nourish without overloading the stomach; and it must bear some relation to the loss of tissue and energy resulting from exercise and other circumstances. It must also to a certain extent be suited to the national and individual tastes and customs, and be directed by the results of experience, both of the physician and the patient.

All these matters should be taken into consideration by the medical director of a sanatorium, who should settle the daily *menu* in consultation with cook or housekeeper, take his chief meals with his patients and notice how they fare, and introduce suitable modifications to suit individual requirements.

The proportions of food stuffs advisable for the majority of early consumptives who are not dyspeptic do not greatly differ from those required in health. As, however, there is a tendency to rapid loss of weight, and the need for rapid constructive metabolism, there should be a relatively large proportion of easily digestible nitrogenous food and fat. Milk, butter and cream are convenient for such purposes, and figure largely in the dietaries of sanatoria abroad. The milk should be obtained from tuberculin-tested cows, as a large proportion of our milch cows are affected with

(41)

tubercle, and in some cases yield contaminated milk.
Where the supply cannot be controlled, the milk should be
boiled before being used; but it is believed by some
authorities that it thereby loses some of its useful properties. However, at Oderberg none but boiled milk is
given, although it comes from carefully tested cows, on
the ground that the patients are more likely to continue
the precaution after their return home. Consumptives in
an early stage are very subject to constipation. For this
and other reasons, ripe fruit and vegetables should figure
largely in their daily diet. Consumptive patients are apt
to be fastidious, and often suffer from loss of appetite, so
that the food must be supplied in a palatable and appetising form, sufficiently flavoured and sufficiently varied to
tempt the palate without offending the eye or upsetting the
digestion. Many common dishes—such as suet puddings
—are digestible or otherwise according to the way in which
they are prepared, so that a good cook is of the greatest
importance to a sanatorium.

The quantity taken at one meal, and the number of meals,
should vary in inverse proportion. A patient who can only
eat a small amount at a time must have frequent meals, while
the patient with a robuster appetite may content himself
with three a day. The meals taken abroad depend so largely
upon national tastes and customs that it is difficult to draw
useful conclusions from them. In France, two substantial
meals—or even one—are the rule among healthy people; in
Germany, three or four are usual; while in the mode of preparing the food the greatest differences exist, even between
different parts of the same country; and what pleases one
patient will disgust another who comes from a different
locality. In some places, cooking is largely done with olive
or poppy oil; in others with butter or various kinds of fat.
The garlic which pleases the Spaniard would nauseate the
average Englishman, who also usually detests the sauerkraut
and vinegar which the German delights in. At some of the
sanatoria—as at Reiboldsgrün—the bulk of the patients

come from the same part of the country ; but in others, as
at Falkenstein, the company is cosmopolitan, so that an
attempt is made to suit diverging tastes by different,
national *menus* on different days. In most of the German
sanatoria two breakfasts are provided ; a mid-day dinner
of three or four courses ; afternoon tea or coffee with milk ;
and an early supper of three or four courses, besides in
some cases milk or soup on rising and at bed time. This is
more than is required by most patients, who would find
three meals enough with two supplementary ones if the
appetite is poor. Debove and others have obtained very
striking results by forcible feeding of consumptives with
concentrated foods ; there is, however, some danger in this
way of increasing the quantity of blood beyond the capacity
of the lungs, and so leading to hæmoptysis, undesirable
fatty changes or overwork of excretory organs.

Physicians are not agreed as regards the use of alcohol.
Whereas Dettweiler recommends full quantities of brandy,
often in a concentrated form, others (such as Liebe at,
Loslau) give none at all excepting in emergencies. It
appears to me that of the two this is the more useful
course, as there is less risk of inducing undesirable habits ;
but I see no reason why alcoholic drinks should not be
used medicinally or dietetically, in reasonable quantities,
and properly diluted, to improve digestion or for other
definite purposes.

A great authority used to teach his students that we
should treat the sick man rather than the disease : and in no
department of medical treatment is it more important to take
into account individual tastes and peculiarities than in diet-
ing consumptive patients. Many of these will of course take
almost anything put before them, but a large minority are
exceedingly dainty and capricious ; and at critical stages will
have to be greatly humoured in the choice of food in order
to induce them to take sufficient for their needs. Sabourin [1]

---

[1] *Traitement Rationnel de la Phtisie*, p. 93. Paris, 1896.

mentions a patient who for a time would scarcely touch any
food but eggs, but partook freely of these in increasing numbers until he was taking from eighteen to twenty-four per
day; and this strange diet sustained him through a critical
period of his illness until he could digest a little bread.
Another patient who could not take milk was able to
digest large quantities of raw meat and alcohol, and made
a good recovery. It is of course most important to attend
to the digestive functions both medicinally and dietetically
—as indeed to the state of all the mucous membranes—but
it is unnecessary to enlarge here upon this topic. Large
quantities of whey seem useful in certain cases in assisting
renal excretion (see p. 171). Febrile patients will need a
simple and digestible diet consisting largely of milk; in
diarrhœa all laxative articles of food must be omitted,
and farinaceous milk foods given together with astringents
and intestinal antiseptics. Where digestion is feeble, pre-
digested foods will often be useful. The diet in a sanatorium in fact will have to range from the "fever diet" of
a hospital to the elaborate dietary of a high-class hotel.

CHAPTER XII.

THERE are on the continent and in America sanatoria for
every grade of society, from those who pay nothing to
those who pay from £4 to £7 per week. They are divisible
into three classes, according as they are intended for those
with means, those of moderate incomes, and the poor. Some
of the cheaper sanatoria are for the " working classes," while
others are for the poorer members of the middle classes,
including teachers, clerks, struggling professional men of
various kinds, and the like. These are really even worse off
than the working classes ; for their incomes are dependent
on their own exertions and on their keeping up appearances,
so that their net income is often lower than that of the
workman. Moreover, they are not rightly eligible for
hospital treatment or for many of the convalescent houses,
while their own domestic surroundings are as little suited
to hygienic treatment as those of the mechanic.

Sanatoria intended for the poorer classes are usually
somewhat different from those where higher charges are
expected. Their rooms are less luxuriously furnished ; the
food is somewhat plainer and less *recherché ;* more than
one patient are often put into the same bedroom ; and a
certain amount of the lighter work is expected to be done
by those patients who are fit for it.

At German sanatoria for the poorer classes, from one to
four patients are usually put into each bedroom. At Ruppertshain as many as six are put into some of the rooms ;

(45)

and at Albertsberg there are dormitories for ten together. This is not a good plan, and should not be encouraged. It is much more difficult to resist the overcrowding of rooms with several beds than of those with one. The larger bedrooms at Ruppertshain were originally intended for five beds, but owing to the numerous applications for admission, an extra bed was put into each. Consumptive patients, whether rich or poor, should each have his own bedroom, just as they do at Ventnor.

No more than the absolutely necessary furniture should be allowed in each room. Each patient usually has his own compartment of a large movable cupboard, for clothes and other small articles. This may be placed in the corridor, so as not to encroach on the cubic space of the bedroom. If lavatories are placed near the bedrooms, as is often done abroad, this will further diminish the necessary bedroom furniture. Enamelled iron utensils should as far as possible be used instead of crockery. There should be a large room near the matron's room for linen and glass, of which a double or treble supply should be provided. A small room is also needed for soap, candles, brushes, blacking, etc. All boxes and trunks should be kept in a box-room, outdoor clothes in a cloak-room, and boots not actually in use in a boot-room ; and all these may be placed in one of the less valuable parts of the building—basement or north side, according to the particular plan adopted. At Oderberg the dining tables are covered with American cloth fastened with a fillet of wood, as this greatly aids in keeping the tables clean. The fillets should, however, be placed on the under side of the table top, which has not been done there.

It is a disputed point whether consumptive patients should do any work or not in the sanatorium. Léon Petit is of opinion that they should not do so : Penzoldt was of opinion that they might safely do light work under medical supervision ; and as a matter of fact in most German sanatoria of this class they are expected to make their beds, keep their rooms tidy, open their windows, clean their own boots, and see to the removal of closet pails where the dry system is in force. At Oderberg Sanatorium each section of the house chooses an overman who is responsible for the performance of these lighter tasks, and for the keeping of various rules. At the Basel Sanatorium at Davos the patients are only too glad to help in the household work ; the women help to prepare vegetables for the dinner and the like ; but it has been found more difficult to obtain suitable employment for the men, who have consequently required more entertainment in the shape of games. Penzoldt sent some patients to the Reiboldsgrün Sanatorium, who were received free of charge on condition that they did light out-of-door work. They are said to have greatly improved in health ; and there was no evidence that their work did them any harm.[1] There are, of course, times when absolute rest is required ; but at other stages light out-of-door work might very well take the place of some of the graduated exercises, and would be better for the patients than indoor work or than idleness and *ennui*. Convalescent patients who had been unsuitably employed before admission might very well devote the last few weeks of their stay to the acquisition of a more suitable employment ; and even if this necessitated the presence of technical instructors it would be a most remunerative outlay in the long run.[2] It is right to regard such matters from a public as well as an individual standpoint ; and every man saved from becoming a pauper or from sinking to a state of invalidism is a gain to the State.

SANATORIA FOR CHILDREN

also require special organisation, as their inmates need more supervision, and are with greater difficulty kept quiet, unless they are gravely affected. Moreover, systematic schooling is in their case essential. The French sanatoria for tuberculous children are a worthy model for imitation (see p. 139).

[1] Penzoldt, " Behandlung der Lungentuberculose," in *Handbuch der speciellen Therapie der inneren Krankheiten.*
[2] See also p. 211.

CHAPTER XIII.

THE RESULTS OF SANATORIUM TREATMENT.

THE statistics which have been published concerning the results of treatment in sanatoria for consumptives are in one sense eminently satisfactory, as they show a very large proportion of apparent recoveries. It must however be remembered that such statistics are often misleading, owing to want of uniformity in the patients received, in the methods of treatment, and in the assessment of the results.

In the first place, the material dealt with is often very different in different places. Some few sanatoria accept all stages of consumption, and only exclude the obviously dying. But most of them only receive such cases as are deemed likely to be benefited; and the standard will necessarily vary according to the views of the medical officers. At Reiboldsgrün patients are kept under observation for a fortnight, and if there is not distinct improvement within that period, they are sent elsewhere. At many other sanatoria more patience is exhibited, as improvement may set in after a much longer period of treatment. The results will also depend upon the social class and antecedents of the patients; and this should always be taken into account. In America it is stated that the poorer patients apply in an earlier and more curable stage of illness than their wealthier brethren; besides which there may be a greater difference in one class than in the other between their food and general conditions of life at home and in the sanatorium. Corresponding with these differences, it is found that the statistics from sanatoria for the poor are slightly more favourable than those from establishments for the more wealthy classes in similar climates. Solly[1] believes his statistics prove that those who are intelligent and well-educated improve more readily than the ignorant or careless, which is extremely probable.

Secondly, the treatment at sanatoria, being made up of a number of separate factors, necessarily varies in different establishments, greater prominence being given to individual factors in one place than in another. In Germany hydropathy is much resorted to; in some American sanatoria it is not used at all. In some places patients take exercise even if they are slightly feverish; in others they are maintained at absolute rest not only in case of fever, but for a variety of other reasons. In some sanatoria patients are carefully kept under medical supervision ; whereas in others much more liberty is allowed, which is not always wisely used. Some establishments are incomplete in their arrangements, or receive a mixed class of patients—asthmatics, bronchitic subjects, and those with heart disease, as well as consumptives, so that supervision is more difficult and treatment less effective. Sanatorium treatment is further complicated by questions of climate and altitude, which of themselves are already sufficiently complex. Such diversity will in future give opportunities for the discovery of many important facts; but so long as the various details are imperfectly recorded, it only adds to the difficulty of drawing any reliable conclusions. The duration of treatment is also most variable : in some cases averaging two or three months, while in others it usually lasts for eight or nine months or even longer.

Finally, the personal equation is again involved in the statement of results, which are differently classified in different places, while the true results can only be obtained by special inquiries instituted some time after the departure of the patient. At some sanatoria the patients who leave are classed as cured, nearly cured, improved, stationary, and worse ; in others as better, stationary, and worse; in yet

[1] *Medical Climatology*, p. 123.

others the results are classed as very good, good, fair, and
bad. A very complete system of classification is adopted
by the Hanseatic Sickness and Old Age Insurance Co.,
under the direction of Dr. Gebhard. The results are
mainly classified under three heads, according to the local
signs, the general condition, and the capacity for work.
Under the *first* head five degrees are recognised, according
as the local signs, (1) originally slight, have disappeared, or
(2) remained stationary, or (3) originally more pronounced,
have diminished, (4) remained the same, or (5) increased.
Under the *second* heading four degrees are recognised : (*a*)
much improved, (*b*) improved, (*c*) stationary, or (*d*) worse.
Under the *third* heading four degrees are recognised,
according as the patient leaves the institution (i.) with full
working capacity likely to be maintained, (ii.) full but
probably temporary working capacity, (iii.) conditional
working capacity, or (iv.) none at all. Under each heading
it is also stated if the results are not evident or not
recorded, or if the patient died. In another table the
effect on the patient's weight is recorded. In other tables
the results are classified according to the original extent of
the disease into seven groups : (1) catarrh of one apex, (2) of
both, (3) extensive catarrh, (4) slight infiltration of one
apex, (5) of both, (6) moderately advanced infiltration, (7)
far advanced infiltration. Another table gives the results
in four groups, according to the general condition on
admission. Yet others show the influence of age and sex,
duration of the illness before treatment, presence of complications, and of inherited tendency to consumption. The
permanence of working capacity in those who regained it
has also been inquired into. Such tables are too elaborate
to reproduce in detail, but they are most valuable for
reference and comparison ; and if similar ones were
prepared by our chest hospitals, and by other institutions
which receive numbers of consumptive patients, the comparison would be most instructive. In the Hanseatic
Company's report for 1897 the results are given in 1541
cases treated from 1893 to 1897, as follows :—

Locally, improvement (1 and 3) took place in 58·1 $\%$
Generally, ,, (*a* and *b*) ,, ,, 85·5 $\%$
Full recovery of working capacity (i. and ii.) resulted in 71·8.

The permanence of results could only be ascertained in 1073 cases ; out of this number 65 $\%$ were still fit for work at the end of 1897. In 77·3 $\%$ of those received up to the end of 1896 the disease had not advanced. In some of these (10·7 $\%$) no abnormal physical signs were discoverable ; in 26 $\%$ local improvement was noted since leaving the sanatorium ; in 40·6 $\%$ no local change. Of those treated in 1897 the results were good even when the general health was decidedly affected at the time of entry, 50 $\%$ of such patients showing improvement (1 and 3) in the local conditions, and 73 $\%$ an improvement in general health. Most of the patients were treated at Oderberg, St. Andreasberg, or Altenbrak.

Less elaborate statistics have been published of the results at many of the German sanatoria.

At the Rehburg Sanatorium (see p. 246) of the Bremen Society, 334 patients were treated from 1st June, 1893, to the end of 1896. Of these, thirty-seven were not certainly phthisical, and are excluded from the statistics. Of the remaining 297, there were on entry :—

71 = 23·9 $\%$ slightly affected.
97 = 32·7 $\%$ moderately affected.
129 = 43·4 $\%$ seriously affected.

The results were as follows :—

	Better.	Unchanged.	Worse.
General condition	253 = 85·2 $\%$	20 = 6·7 $\%$	24 = 8·1 $\%$
Local condition	63 = 21·2 $\%$	194 = 65·3 $\%$	40 = 13·5 $\%$.

Of the more serious cases 75·2 $\%$ increased in weight, and of the slighter cases, 81·8 $\%$.

In the report for 1897, the results as to working capacity ·in ninety-four undoubtedly tuberculous cases are worth quoting :—

Restored to full, and probably permanent, working capacity 38·3 $\%$
Restored to full, but probably temporary, working capacity 35·1 $\%$
Restored to conditional working capacity . . . 16·0 $\%$
Unfit for work . . . . 10·6 $\%$.

Beaulavon[1] gives the results on working capacity of patients treated during 1894 and 1895 at the same sanatorium. From this it appears that out of 170 cases, 37 $\%$ were restored to their full working capacity, and 9·4 $\%$ more were able to do light work. In the first stage, out of 53 patients, 81·1 $\%$ were restored to full working capacity; in the second stage, out of 89 patients, only 22·5 $\%$; in the third stage none were restored to full, and only 28·6 $\%$ to conditional, working capacity. The influence of inheritance on the results seems to have been inappreciable.

At Dr. Weicker's Krankenheim in Görbersdorf (see p. 209), 185 patients completed their treatment in 1896. Out of these—

130 = 70·3 $\%$ had regained their working capacity.
18 = 9·7 $\%$ were capable of light work.
22 = 11·9 $\%$ were better but not fit for work.
15 = 8·1 $\%$ were no better.

Manasse[2] has published the results of 5032 patients treated from 1876 to 1886 inclusive, at the Brehmer Sanatorium at Görbersdorf (see p. 150).

Stage of Disease.	No.	Cured.	Nearly Cured.	Total Improved.
I.	1390 = 27·6 $\%$	387 = 27·8 $\%$	430 = 31 $\%$	817 = 58·8 $\%$
II.	2225 = 44·2 $\%$	152 = 6·8 $\%$	325 = 14·6 $\%$	477 = 21·4 $\%$
III.	1517 = 28·2 $\%$	12 = 0·8 $\%$	33 = 2·3 $\%$	45 = 3·1 $\%$
	5032	551 = 11 $\%$	788 = 15·6 $\%$	1339 = 26·6 $\%$

From Dettweiler's statistics of 1022 patients treated at the Falkenstein Sanatorium, it appears that 13·2 $\%$ were apparently cured, 11 $\%$ nearly cured, or a total of 24·2 $\%$ greatly improved.

Statistics of 2000 patients treated at Reiboldsgrün (see p. 160) under Dr. Driver showed that 13·6 $\%$ were cured, 28 $\%$ greatly improved, 28·6 $\%$ improved, 25·2 $\%$ un-

[1] *Rev. de la Tuberculose,* April, 1897.

[2] *Die Heilung der Lungentub. durch diätetisch hygienische Behandl. in Anstalten und Kurorten,* Berlin, 1891.

improved, 4·5 °/₀ died.[1] From the two most recent reports
it appears that in 1896, of 349 patients who left the institution, 263 (or nearly 76 °/₀) were improved; in 1897, of
366 patients, 295 (or 80 °/₀) were improved. These figures
include every degree of improvement. It should be remembered that owing to Dr. Wolff-Immermann's system
of selection, a relatively favourable material is dealt with
at this sanatorium. On the other hand, the average duration of treatment is only sixty-six to seventy days, which
is less than at most sanatoria. The following tables show
the results at some of the foreign sanatoria:—

PAY SANATORIA.

Name.	No. of Cases.	Stages.	Appar- ently Cured.	Nearly Cured.	Im- proved.	Total Pro- portion Im- proved.	Authority.
Brehmer . . .	5032	all	11	15·6	...	...	Manasse.
Do., excl. 3rd st.	3615	1 & 2	14·9	20·9	...	...	Do.
Do.	...	all		25	50-55	75-80	Achtermann (Ransome).
Do.	...	...		25·1	60·9	86	Kobert (Hohe).
Falkenstein . .	1022	?1 & 2	13·2	11	...	...	Dettweiler
Do. . .	...	...	14	14	45	73	Do. & Hess (Ransome).
Do. . .	...	...	14-15	nearly as many	...	...	Hess, *Practitioner*, Nov 1897
Reiboldsgrün .	2000	?1 & 2	13·6	28	28·6	70·2	Driver.
Do. .	715	?1 & 2	...	...	...	78	Wolff-Immermann.
Hohenhonnef .	...	?1 & 2	14·5	28·9	...	69	Meissen (Ransome).
Nordrach . .	...	all		30	65	95	Walther do.
Römpler . .	...	...	...	...	...	75	Römpler (Hohe).
Weicker . .	...	...	...	...	...	84·8	Weicker do.
Schömberg .	...	...	...	...	...	82·9	Baudach do.
St. Blasien .	...	...	...	...	...	84	Sander do.
Canigou . .	...	...	...	43·8	...	...	Sabourin (Ransome).
Do. . .	...	...	22-23	...	40-50	62-73	Giresse (private letter).
Davos-Turban .	...	no advd.	20	30	40	90	Turban.
Do. .	...	do.		40	40	80	Do. (Ransome).
Arosa . . .	259	few advd.	...	...	...	82	Jacobi.
Leysin . .	79	all	12·7	...	59·5	72·2	Burnier (Montmeylian).
Do., 1st stage .	15	1st	53·3	...	33·3	86·6	Do. do.
Do., 2nd stage .	22	2nd	9	...	86·3	95·3	Do. do.
Do., 3rd stage .	42	3rd	...	...	50	50	Do. do.
Do. . . .	...	all	12·5	...	56·5	69	Exchaquet.
Winyah—							
Early stage . .	...	...	81	...	19	100	v. Ruck (private letter).
More advanced .	...	...	35	...	23	58	Do. do.
Advanced, but still in fair gen- eral condition .	...	...		...	24	33	Do. do.
All stages . .	...	all	9	22·6	42·5	65·1	Do. (Ransome).
Hygeia . . .	...	all	22·5	...	46·25	68·75	A. C. Klebs (private letter).

[1] "Volkssanatorien für Lungenkranke," *Deutsche Med. Zeit.*, 1890.

SANATORIA FOR THE POORER CLASSES.

Name.	No. of Cases.	Stages.	Appar- ently Cured.	Nearly Cured.	Greatly Im- proved.	Im- proved.	Total Propor- tion Im- proved.	Authority.
Falkenstein, for the poor }	...		13	...	...	77	90	{ Dettweiler & Nahm (Ransome).
Ruppertshain	313		...	...	...	...	77·6	1895-6. Nahm.
Malchow . .	...		...	...	43·7	40·3	84	Reuter (Hohe).
Rehburg . .	...		...	...	59·5	25·2	84·7	Michaelis (Hohe).
Schömberg .	...		...	... -	60·9	33·3	94·2	Baudach do.
Weicker . Krankenheim }	...		...		70	22	92	Weicker do.
Jonsdorf . .	...		...	...	20	73	93	Toop do.
Königsberg .	...		...	...	72	13	85	Andræ do.
Grabowsee .	219		...	14·1	...	64·4	78·5	Brecke,1896-7(Liebe).
Blankenfelde	239		...	...	58·1	38·9	97	Ellerhorst, 1896 do.
Hanseatic Insur. Co. }	1541	{ slightly affected 30·9 % }	...			...	...	Report for 1897.
Local . .	...		...	...	...	...	58·1	Do.
General . .	...		...	...	...	...	85·5	Do.
Halila Alexander }	300	{ 60 % 2nd } 27			...	43·7	70·7	Gabrilovitsch,5 years.
Davos (Basel)	185	{ 20 % severely affected } 25·4		42·3		23·1	90·8	{ Kündig, Report for 1897.
Adirondack	{ mean of 10 years }		...	20-25	20-30	...	...	Trudeau (Knopf).
Do.	105		...	21·9	35·2	...	20	77·1	{ Do., 12th Annual Report.

Solly[1] discusses the statistics from various parts of the
world, and concludes that the percentage of improvement
is as follows :—

	All Stages.	First Stage.	Second and Third Stages.
Lowland climates .	58	71	28
Sanatoria . .	63	95	58
Highland climates	76	89	63

As the statistics of sanatorium cases were exclusively
from those in non-Alpine climates, this is striking testi-
mony as to the value of sanatorium treatment.   Generally
speaking, one may say that from one-fourth to one-third of
the patients treated in sanatoria are practically cured, or a
still greater proportion if they are treated in an early stage.

¹ *Medical Climatology*, p. 141.

Probably systematic and prolonged treatment from an early stage would restore to health from one-half to two-thirds of our consumptive patients, even without the advantage of an Alpine or other high altitude station. Unfortunately, it is quite out of the question to expect patients to submit to more than a few months' treatment in a sanatorium, so that we must trust to the educational influence of the sanatorium to complete the recovery of those treated in it. Improvement frequently—perhaps usually—continues after patients have left the sanatorium, if only the conditions of life are fairly satisfactory.

The good results of treatment are permanent in a large proportion of cases. Special inquiries were made in the year 1890 to ascertain how many of those treated at the Brehmer Sanatorium continued in good health. In five cases the cure had lasted from twenty to twenty-nine years ; in fifty-two cases, from twelve to twenty-one years ; and in thirty-eight cases, from seven to twelve years. Of forty patients discharged from the sanatorium in 1876 as cured or nearly cured, of whom particulars could be obtained, there were still twenty-five living in good health, while one suffered from fibroid phthisis, one had died four years previously from phthisis, and thirteen others had died from unknown causes.[1] Similar investigations by Dettweiler, at Falkenstein, in 1886, led to the discovery that seventy-two out of ninety-nine patients who had left the institution as " cured " were still living from three to nine years after in perfect health ; in fifteen cases there had been relapse, although twelve out of these fifteen subsequently recovered.[2] At St. Blasien, Dr. Haufe inquired in 1891 after all those who had been treated there from 1878 to 1889. Forty-six did not answer; five were dead ; twelve had had a relapse after an interval of three

[1] Wolff und Saugmann, *Ueber die dauernde Heilung der Tuberculose*, Wiesbaden, 1891.

[2] Dettweiler, *Bericht über 72 seit 3-9 Jahren in Falkenstein völlig geheilte Fälle von Lungenschwindsucht*, Frankfurt, 1886.

to six years; 201 who left from two to twelve years
previously were at work without difficulty, although they
continued to cough; seventy-two were apparently quite
cured, and had been so for three to twelve years. Amongst
the latter were six officers who for several years (one for
six years) had done their work well without interruption.
Some had originally had symptoms of acute phthisis; the
others had been long tubercular and had repeatedly coughed
up blood. Of the seventy-two, twenty-one were in the
third period: these had lost their bacilli in the sanatorium
and increased in weight. Assuming that the non-replies
were worse or dead, 21·4 per cent. might be regarded as
cured, and 59·8 per cent. more as in good health.[1] Knopf
states[2] that Dr. v. Ruck, of Winyah Sanitarium, wrote to
605 patients who had left this institution from one to three
years previously. He received 457 replies, which showed
that 14·6 per cent. were absolutely cured; in 15·3 per cent.
more the disease had made no further progress; and 56·4
per cent. were better than when they left the sanatorium,
or a total of 86·4 per cent. better or cured, 13·6 per cent.
worse or dead. If all the non-replies were dead, these
figures would be reduced to 65·3 and 34·7 per cent. respectively. The results obtained by Dr. Gebhard of the
Hanseatic Insurance Co. have already been quoted.

Every victim of unarrested phthisis is a possible focus for
the dissemination of the disease. He is also a burden to his
family, or to the State, or both, for months or years, and, if
a bread-winner, plunges those dependent upon him into
serious pecuniary and social difficulties. Remembering these
facts, and remembering also that, in even the best managed
hospital, the chances of recovery for the average consumptive are much smaller than at a sanatorium, we may more
easily realise the important services which these institutions are able to perform. From the statistics of Brompton
Hospital, it appears that the percentage of improvement is

---

[1] Manasse, loc. cit.   [2] Les Sanatoria, Paris, 1895.

only 20 to 30, as compared with 50 to 90 at sanatoria ; and
making every allowance for the possibility that serious or
unfavourable cases are treated in larger proportion at the
hospitals, it must be evident that they are quite unable to
play the part of sanatoria. Weeks and months often go by
while the poor consumptive waits for admission to a chest
hospital. Were sanatoria established in the numerous suitable localities in England, the hospitals would be relieved
and the patients cured in far larger numbers.

CHAPTER XIV.

THE size of a sanatorium depends partly on the number
which can properly be supervised by one medical officer,
partly on financial considerations. A small sanatorium is
relatively expensive to erect and maintain, as the administrative parts for a dozen patients would almost suffice
for three dozen, and the expenses per head rapidly diminish
with increase of numbers. A small sanatorium is also apt
to be dull, as there is less probability of patients finding
congenial companions, as well as for other reasons.

On the other hand, no medical officer can properly manage
more than thirty or forty patients unaided, so that with
two medical officers sixty or eighty is probably the best
number, as recommended by Léon Petit.[1] Some of the
foreign sanatoria far surpass this modest number. The
Brehmer Sanatorium at Görbersdorf has now 300 beds
with eight physicians. Most of the newer sanatoria for
the working classes in Germany are being built with ninety
or 100 beds.

The nursing staff in many foreign sanatoria is somewhat
insufficient according to English notions, even taking into
account the stage of illness for which the sanatorium is
designed. At the Brehmer Sanatorium there are nine male
and female nurses, including three bath attendants. At
Falkenstein there is no proper accommodation for the nurse
if the patient is confined to his bedroom; and trained

[1] *Loc. cit.*, p. 52.

nurses are only brought in for emergencies, the usual
attendants not being hospital-trained.   At Hohenhonnef
one male and one female nurse are provided for each floor,
or about one to every fifteen patients, which is probably
ample.   At Reiboldsgrün, with over 100 beds, one trained
nurse is kept ; but the persistently febrile patients are not
retained in the establishment, as Dr. Wolff-Immermann
believes that a patient who does not improve in the first
fortnight is unlikely to do so at all in that place.

Sanatoria for paying patients naturally have a larger
staff than those for the poorer classes.   Hohenhonnef, with
a *personnel* of seventy-nine, has probably the largest staff
of any in comparison with the number of its patients.   Of
sanatoria for the less wealthy classes, Oderberg may be
taken as an example.   This is an institution with 120
beds, under an inspector, and has four male nurses, a
female cook and three kitchenmaids, a machine tender
and his mate, a heater, a steward and his wife, coachman, and messenger.   Including the outside hands, there
are altogether twenty-five on the staff.

The Felixstift, at St. Andreasberg, which has at present
thirty-two beds, has only a matron, three maids and a
steward.   Eventually it will have forty beds, and a somewhat larger staff.   At Stiege, with fifty-eight patients and
eventually seventy-two beds, the staff number nine, including the steward.   The Rehburg Sanatorium of the
Bremen Insurance Company, which has thirty beds, is
managed by a matron, with a cook, kitchenmaid, housemaid and man-servant under her.   The Sülzhayn Sanatorium, which when finished will have 120 beds, is to be
managed by a matron, engineer, female cook, two kitchenmaids, two male nurses and an undetermined number of
female nurses, messenger, and night watchman.   No inspector will be appointed.   The daily cleaning is to be
done by an additional staff of women from outside.

As in a hospital, the administrative work of a sanatorium
should be in the hands of a layman, manager, steward or

managing director, or in a small sanatorium a matron or housekeeper; but it is essential that he or she should be subordinated to the chief medical officer, with an appeal if necessary to the managing board or committee. There is a feeling in this country amongst the medical profession against one of their number being the owner of any kind of medical institution—whether private hospital, nursing home, hydropathic establishment, asylum or sanatorium. This feeling is, in my opinion, well founded ; and while it is only right and proper that the medical officer should receive adequate remuneration, this should not be directly in the shape of patients' payments, nor should he be directly responsible for the finances of the establishment.

The admission of patients may be managed somewhat after the method adopted by the Basel Sanatorium at Davos. There is a medical board in the town of Basel, to whom applications are sent, signed by the family medical attendant, and stating the stage of the illness and the condition of the patient. The latter then calls on a member of the medical board at an appointed hour, and if approved starts on his journey to the sanatorium. The resident medical officer, however, has power to send home any patient who appears to be unsuitable for treatment.

Patients enter for at least thirteen weeks at the Basel Sanatorium. Three months is a very common duration of treatment abroad, although most medical officers of sanatoria regard this as barely sufficient to train the patient in necessary hygienic methods and to put him on the road to recovery. At Canigou patients who begin to be home-sick are sent home for a time, to return again for further courses of treatment. Dr. Giresse regards this interruption as a concession to human frailty rather than as medically desirable. In some sanatoria quite a different system prevails, the patient staying until he is apparently cured, irrespective of how long this may require. Thus, at Nordrach patients stay until for about twenty examinations no tubercle bacilli have been found in their sputa, and until the injec-

tion of their sputa into a guinea pig (which is next done)
does not cause tuberculosis in the animal. At Davos and
other Alpine sanatoria at least two winters and one summer
are advised ; in Colorado, too, at least two years' residence
is considered necessary. At the Adirondack Cottage Sanitarium patients stay until they are apparently cured ; the
average stay being six months, and one year nominally the
limit.

It seems to me that it depends on circumstances what
course should be pursued. Where the patient comes
from a suitable home and district, and can carry out the
treatment in a modified form after his return home, this
may be permitted after a few months if he has made progress; but if he relapses he should be promptly sent back
to the sanatorium. Where the locality or the domestic
circumstances are unsuitable, a longer residence—if possible
until apparent recovery—is advisable. In every case the
course should be pursued which promises the best medical
results ; pecuniary considerations are of very subordinate
importance in dealing with a dangerous disease. And if
the patient's finances will not stand the strain, some means
should be devised whereby he may be assisted. A man
who is held to ransom by brigands, or who is drowning at
sea, does not stop to count the cost, but leaves his money
to save his life ; a consumptive patient should do the same.

The sanatorium, however, should not be expected to receive hopeless cases ; these, if not treated at home, should
enter a different class of institution—the home or refuge for
incurables. There are several such institutions in America,
France and Switzerland ; possibly also in Germany, and
several in England. More of these refuges are needed in
this country for the poor, who are not welcome in either
general or chest hospitals, and usually gravitate to the
poor-law infirmaries. In the course of time, with a more
perfect organisation, the number of such cases will diminish,
but in the meantime institutions for their reception are as
necessary as sanatoria, both for the sake of the public and

for that of the patient. The chest hospitals will remain to do a most valuable work, for which they are fitted, while they give up functions for which they are utterly unsuited. Acute cases and those presenting intercurrent inflammatory attacks will find a welcome refuge in these institutions, to be passed on when convalescent to a sanatorium. The value of a chest hospital is to a great extent sacrificed when patients have to wait for weeks before being admitted ; and this drawback will disappear when auxiliary institutions, such as sanatoria, have been provided on a sufficient scale.

Patients who return convalescent from a sanatorium to an unsuitable neighbourhood or to unsuitable homes and occupations, point to the need of further philanthropic organisations. There is no reason why model sanitary colonies should not be established in suitable districts, where cured consumptives could lead a healthy out-door life, more or less under medical guidance and with the provision of specially airy dwellings and workshops. This is practically what has been done in Colorado, where the city of Denver is largely peopled by such patients ; a company was projected (perhaps started) in 1891 for the erection of such a village at New Florence near the Gulf of Mexico (see p. 98) ; and there is a similar movement in Germany for the provision of model country dwellings. In France, the convalescents from the Villiers Sanatorium are sent to agricultural colonies (see p. 144) ; and wealthy England should be capable of a similar organisation. But with all these institutions, the sanatorium will remain a most important link in the chain of treatment.

CHAPTER XV.

TREATMENT of consumptives in sanatoria might conceivably be opposed on sentimental, ethical or medical grounds. It is sometimes said that such treatment must be sad for the patients, as well as for those who see them, and that the deprivation of all the delights of home life must render them unhappy. But there is not really much force in these objections. I can testify that, as Léon Petit and others have also remarked, the patients at these sanatoria seem to be by no means unhappy, but rather rejoice in the greater comfort and more steady progress towards recovery which are possible in such establishments. Much time and trouble are expended by the medical officers and others to keep the patients free from ennui ; and enough is going on as a rule to make this an easier task than would perhaps be imagined. As for deprivation of home comforts, this depends largely upon the particular institution. In some sanatoria, where a very strict rule is observed respecting visitors and companions, and where the manners and customs are foreign, an English patient might very possibly feel the loss of home comforts ; but this is certainly not, and need not be, the case everywhere. At Reiboldsgrün a special house has been built in the sanatorium grounds for the reception of friends and relatives of patients, and for convalescents who wish to revisit the place. In this, or some similar way, there should be no difficulty in the patient keeping in touch with his relatives and friends, provided always that these are dis-

(63)

creet and do not interfere with the treatment. Aggregation of patients there must be to a certain extent, in order to obtain the benefits of co-operation ; but isolation is seldom required—far less than during the course of other febrile complaints, as the isolation (so far as it is needed) is for the sake of the patient rather than in order to prevent infection, which is with reasonable precautions quite unlikely to happen. As a matter of fact, life in a properly-managed sanatorium is far more pleasant than in an ordinary health resort for consumptives, where all kinds of cases and all stages of phthisis are received, and medical supervision is necessarily less effective.

It might be argued that the establishment of sanatoria for consumptives damages the prospects of the ordinary medical adviser. I have never yet heard so selfish and short-sighted an argument ; for the medical profession is ever ready to take a wide and enlightened view of whatever concerns the good of its patients, and (as the achievements in sanitary matters amply testify) often promotes measures which are intended to prevent disease. As a matter of fact, sanatoria not merely benefit consumptives, but also benefit their family advisers, inasmuch as the patients' lives are prolonged, while medical and hygienic advice will be needed for some time after leaving the sanatorium. Treatment in such an establishment is of strictly limited duration ; and is intended, not to cure the patient during the time of his stay—for this is often impossible— but rather to put him on the road to recovery. The good obtained in the sanatorium is often very striking ; but the promise of future improvement is even more important. It is only where a patient would have to go to an utterly unsuitable neighbourhood or mode of life that the period of treatment would have to be considerably prolonged ; and in this case a kind of hygienic colony would best meet the difficulty. In any case, the treatment at a sanatorium will only continue until the patient can safely go home with a prospect of further improvement.

There are no purely medical arguments against sanatorium treatment which will bear examination. The danger of infection is far greater in an open health resort. The experience of chest hospitals and sanatoria, both here and abroad, shows that there is absolutely no danger if simple precautions are observed. At the Brompton Hospital careful inquiries, extending over a period of thirty-seven years, were made by the late Dr. Cotton and Dr. Theodore Williams as to possible infection from patients. The old building was very badly ventilated ; but although the foul air from phthisical patients produced attacks of sore throat and erysipelas, it did not lead to spread of tubercular disease amongst the healthy attendants. None of the resident medical officers, matrons, gallery maids, porters or secretaries and clerks became phthisical, although most of these were brought into frequent contact with the patients. Out of about 150 house physicians only one appeared to have contracted the disease in hospital ; out of 101 nurses three died of consumption after leaving the institution, but in only one did the disease show itself while in hospital. Of twenty-two dispensers three died of phthisis, one while in the building: and two of the dispensers held office for twenty years.[1] Similar investigations by Heron at the City of London Hospital for Diseases of the Chest also failed to prove infection among the attendants.[2] No case has ever been reported from any modern chest hospital which takes even elementary precautions concerning the sputa. Aufrecht states that at the hospital of Magdeburg-Altstadt 34,560 patients were received during a period of seventeen and a quarter years, of whom 3820 were phthisical, mostly in an advanced stage ; but none of the other patients, and none of the large nursing staff, became consumptive. Two tabetic patients and four with multiple sclerosis remained from three to eight years side by side with consumptives,

[1] See Pollock, *Practitioner*, June, 1898 ; Wilson Fox, *Diseases of the Lungs and Pleura*, London, 1891.
[2] *Lancet*, 6th Jan., 1894.

5

and another with adherent pericardium became intimate with his phthisical neighbour, without any becoming infected.[1] At the meeting of the American Climatological Association, in May, 1896, Dr. V. Y. Bowditch, of Boston, said : " I wish to refute the statements that properly-regulated consumptives' hospitals are a source of danger to the community, when I believe them to be exactly the opposite, as shown by statistics ". This opinion will be accepted by all practical physicians ; but in view of the scare which appears to be arising among certain sections of the laity the matter needs to be called attention to.

Dr. I. H. Hance, assistant to Dr. Trudeau at the Adirondack Cottage Sanitarium, proved by the inoculation of guinea pigs that sixteen out of seventeen cottages inhabited there by consumptives for so long a period as ten years were absolutely free from infectious material. In the exceptional cottage the patient had disobeyed instructions and expectorated wherever convenient. In a further investigation Dr. Hance took dust from tenement houses containing consumptives. Where instructions had been followed no guinea pigs suffered; while in the dirtier tenements two out of three were found infected. In street cars one out of every five were found to be dangerous. In the two hospitals, Bellevue and Charity, no infectious dust was discovered, excepting in the out-patient room at Bellevue. In the Winyah Sanatorium none was found.[2]

Römpler investigated the mortality from consumption from 1790 to 1889 in the village of Görbersdorf, which is close to several large sanatoria, with an aggregate of 500 to 600 beds. Before the establishment of the oldest sanatorium the deaths from consumption in the village were at the rate of 0·83 per annum ; whereas since that time the

[1] *Zur Verhütung und Heilung der chronischen Lungentuberkulose,* Vienna, 1898, quoted by Römpler, *Deutsche Medizinal Zeitung,* 1898, No. 35.

[2] *N. Y. Med. Rec.,* 28th Dec., 1895.

OBJECTIONS ANSWERED.

67

rate was 0·47 ; and yet the population had doubled in
twenty-five years, and in forty years some 25,000 consumptives had been treated in the different sanatoria.[1] Nahm,
who made similar investigations in the village of Falkenstein, obtained corresponding results. During the twenty
years preceding the establishment of Dettweiler's institution
an average of 4 per 1000 of the inhabitants died annually of
consumption. After the sanatorium was opened the average
annual mortality from this disease fell to 2·4 per 1000.[2]
Dr. Römpler has had five servants for twenty-three years at
his large sanatorium, and seven more than five years, out of
twenty-three in his present staff; not one of these twenty-
three are consumptive, nor any whom he can trace who have
been in his employ, although they necessarily come freely
into contact with the patients.[3] Dr. Achtermann, for many
years connected with the Brehmer Sanatorium, and now at
Laubbach, states that he was for years in the habit of test-
ing by inoculation the dust from the corridors, saloons,
W.C.s, and patients' rooms. Only once did he find evidence of the existence of tubercle bacilli, on a washing
board where a spitcup had stood.[4] Ransome states that
sputum, which retained its virulence for several months in
a poor cottage in Ancoats, entirely lost its power of communicating the disease to guinea pigs by inoculation when
freely exposed to the air and light in a consumption hospital, and in a well-lighted, well-ventilated, and well-drained
house.[5] Moreover, in another series of experiments, sputa
as well as pure cultivations lost their power for evil on exposure to air and light for two days, or to bright sunshine
for one hour.[6]

[1] *Beiträge zur Lehre von der chronischen Lungenschwindsucht*, Berlin,
1892. See also Knopf, *loc. cit.*
[2] *Münch. Med. Wochenschr.*, 1895, No. 40.
[3] *Deutsche Medizinal Zeitung*, 1898, No. 35.
[4] Prospectus of Laubbach Sanatorium.
[5] Ransome, *The Treatment of Phthisis*, London, 1896, p. 37 ; Ransome
and Dreschfeld, *Proc. R. Soc.*, xlix., 66.
[6] Ransome, *ibid.* ; Ransome and Delépine, *Proc. R. Soc.*, lvi., May, 1894.

Even in open health resorts there is very little danger of infection excepting in crowded towns and cities, or where ordinary hygienic precautions are neglected. It is true that at the Riviera the deaths from consumption amongst the native inhabitants are said to have increased; but this is not quite certain, and is in any case capable of explanation in other ways. It is quite possible that the hotter climate may encourage the saprophytic existence of the tubercle bacillus. But the conditions of life in a densely-crowded town are very different from those in a scattered village; and if the facts are correct, they point to the need of more reasonable methods of building, better ventilation, and less overcrowding, together with more systematic precautions concerning cleanliness, disposal of the sputa, and the like. Consumptives should not live in towns, or if compelled to do so should inhabit dry and well-ventilated rooms. If they do so, and refrain from random spitting and other uncleanly ways, they will never be a serious danger to their companions. Ransome states[1] that he has never seen a case of infection in an ordinarily well-ventilated house, and gross neglect of ordinary rules would probably be needful to cause tubercular infection in such a house. Haupt made a careful investigation at Soden, and found that there were fifty-two people between seventy and ninety years old who had lived in thirty-one different houses which they let out in summer to consumptives. Half of them also waited on the invalids, but none became phthisical. Michaelis has practised over thirty years in Bad Rehburg. This bathing resort has existed for over fifty years, and is annually visited by about 500 consumptives. There are at present about 350 inhabitants in sixty-five to seventy houses; and only three consumptives were found amongst them who had been born there, while these had acquired the disease elsewhere.[2] Even amongst married

[1] Ransome, *The Treatment of Phthisis*, London, 1896.

[2] *Monats. f. prakt. Balneol.*, 1897, No. 1; quoted by Unterberger, *St. Pet. M. Woch.*, 1897, n. F. xiv., No. 29.

couples, where one is phthisical it is exceptional to find the other affected. Leudet found only seven out of 112 such cases, and Haupt only 7 per cent. out of 1061 married couples.[1]

It is perfectly certain that under ordinary hygienic conditions the danger of infection from a phthisical patient is purely imaginary. But this should not blind us to the very real danger which exists where hygienic rules are disregarded, and this is in fact a strong argument in favour of hygienic training of consumptives in a sanatorium.

It may be argued that since many patients recover from consumption without being treated in a sanatorium, this is unnecessary. It is perfectly true that such recoveries take place, but they can only be expected in selected cases; and as Prof. v. Ziemssen has said: "The possibility of treatment outside a sanatorium with equally good results cannot be denied, but it requires much more prolonged rest, and much more time on the part of the physician, and has by no means so certain a result".[2]

It is a financial impossibility to provide one or two patients at home at reasonable cost with the systematic treatment they would obtain in sanatoria, and although patients may be able occasionally to do without these aids to recovery, there are stages and forms of the disease in which it would be most unwise to discard them, if they could be found within a reasonable distance from home. Even ordinary treatment at home, if properly carried out, is more expensive than is usually realised; and this is without any special shelters or conveniences for exercise in the open air; often without skilled nursing in case of feverish attacks, night sweats, or sudden hæmoptysis; and without systematic training or graduated exercise, which indeed can scarcely be carried out under the circumstances. Once

---

[1] Unterberger, *loc. cit.*

[2] *Ueber den gegenwärtigen Stand der Behandlung Tuberculöser*, Berlin, 1897, pp. 22, 23.

it is admitted that these measures are of value in restoring
the consumptive to health—and this can scarcely be disputed—it follows that special institutions are needed to
put them within the reach of those who cannot afford a
special establishment of their own.

CHAPTER XVI.

THE COST OF A SANATORIUM.

SANATORIA for consumptives differ enormously in original cost. · Even apart from the value of the land on which they are placed, it is possible to spend much or little in constructing the buildings, according to the design, the natural obstacles to be overcome in laying the foundations, the cost of labour and materials, the luxuriousness of the accommodation, the cost of water supply, lighting and sewerage, and a number of other circumstances. Even among sanatoria for the poorer classes there are considerable differences in the cost per bed; and those for the wealthy present as much diversity as do inns and hotels in various places. So many more sanatoria have been built in Germany than in any other country, that most of the available information concerning the cost of sanatoria comes from that country. It is difficult to obtain authentic information as to the cost of sanatoria for paying patients. Hohenhonnef, which is perhaps the most luxurious of recent establishments, appears to have cost nearly £72,000, or about £660 per bed, without reckoning the ground, but including the recently-built doctor's and director's residences. It is more difficult to estimate the cost of other first-class sanatoria, even where trustworthy figures are available, as the original cost per bed always largely exceeds the final cost after successive additions have been made. Enormous sums have been spent on the Brehmer Sanatorium, as it is situated in an unusually large and fine park, and much money has been lavished on unnecessary

architectural ornamentation, as well as more practical advantages. Judging by a statement by Léon Petit,[1] a
building of the size and character of Falkenstein Sanatorium would cost in France or Germany about £40,000 for
100 beds. Of first-class sanatoria, probably the one at
Nordrach has been the most simply and least expensively
constructed. The abundant water supply there considerably diminishes the cost of electric lighting ; while the
buildings, though carefully finished, are mainly of wood.
Building operations are more expensive in England than
abroad ; but there is reason to believe that a satisfactory
sanatorium for paying patients could be constructed in a
suitable locality for considerably less cost per bed than has
been done at Hohenhonnef.

Turning now to the institutions for the poorer classes,
we find that they cost on the Continent from £120 to over
£300 per bed. Prof. v. Leyden gives £150 as a fair average estimate, or a total cost of £200 to £250; but this only
refers to the larger ones with about 100 patients.[2] Small
sanatoria necessarily cost more per bed than larger ones, as
the administrative portions for a small number of patients
are nearly as costly as those for twice or thrice as many.
Electric lighting, too, is relatively much more expensive for
a small number of rooms ; and the ground required for a
small sanatorium would equally well accommodate a large
one. Sanatoria for both sexes are more expensive than
those for only one, as many parts have to be duplicated.
This, of course, applies much more to a people's sanatorium
than to one for the middle and upper classes. Sanatoria
built in one block are less expensive than those for a
similar number of patients lodged in several smaller buildings. This has already been alluded to in chap. vii. It
is customary in this country to reckon the cost of a solidly-
constructed building at 9d. per cubic foot. Wooden build-

---

[1] *Loc. cit.*, p. 258.
[2] *Ueber den gegenwärtigen Stand der Behandlung Tuberculöser*, Berlin,
1897.

ings and "bungalows" are less expensive, and in most
years would do perfectly well for a sanatorium. A building
for sixty beds would probably cost as much as three separate buildings, each to accommodate fifteen patients.

In planning a sanatorium, the possibility of extension
must be taken into account. At Grabowsee, a large number
of patients have been accommodated in "Döcker'sche Baracken," which are temporary erections constructed of a
sort of painted paper on a wooden skeleton. These are too
cold for use in severe winters such as are common in North
Germany; and are being gradually supplemented by more
solidly-constructed buildings for use all the year round.

It is false economy to erect a sanatorium on a small patch
of ground, unless this is surrounded by public woods or
moors of a suitable kind. It is also usually a mistake to
attempt the transformation of a pre-existing building into
a sanatorium. The prime cost is undoubtedly thereby
often diminished; but the alterations and additions which
are nearly always required, are very apt to swallow up the
resulting economy. According to Kuthy, a house in Berlin,
transformed into a hospital, cost £128 per bed for alterations. Few buildings not of recent construction are satisfactory from a hygienic point of view. Nature's purifying
processes take time to remove the effects of leaky sewers
and drains. Solid foundations and damp-proof courses are
both difficult and expensive to put into an already constructed house. In a sanatorium everything is (or should
be) sacrificed to superabundant ventilation and free access
of sunlight; whereas the average house is constructed with
the idea of making the largest possible number of rooms on
a given space. Many convalescent homes and hydropathic
establishments would, however, have to be excepted from
this statement, and, if suitably situated, might be usefully
converted into sanatoria. A hospital in a town costs more
than a sanatorium of the same size in the country. The
Basel-town Hospital is estimated to cost £400 per bed,
whereas the Basel-town Sanatorium at Davos has been

erected at a cost of £220 per bed.[1] In England the cost of
erecting hospitals amounts to £450 or £500 per bed (Burdett). A very good popular sanatorium could be erected
for much less than this in a suitable country place, while
the site would be far less costly for the latter. Kuthy [2] gives
the cost of the Ruppertshain Sanatorium as follows :—

		Marks.
Mason's work, iron work, etc.	. .	100,000
Carpenter's work and flooring	. .	35,000
Roof	.	4,500
Tinsmith's work	.	1,500
Windows, doors, etc. .	. . .	22,000
Varnishing	. .	7,000
Heating apparatus	. . . .	17,000
Water supply	. . . .	10,000
Trees, etc.	. .	1,500
Entrance, paths, etc. .	. . .	1,500
Total	. . .	200,000, or about £10,000.

Add to this 45 to 50 mks. per bed for furnishing. The
building was originally intended for 75 beds, but owing to
the demand for admission 88 patients are actually admitted,
an equal number of either sex. Kuthy states that the
total cost was 210,000 mks. (probably with the ground),
exclusive of furnishing. This is equivalent to £140 per
bed for 75 beds, or nearly £120 for 88 beds. According to
Léon Petit [3] the institution cost 325,000 frs., or 4333 frs. per
bed for 75 beds, including furniture. Kuthy also gives the
estimate for a sanatorium of 35 beds for the city of Worms,
which may be condensed as follows :—

		Marks.
Ground	. .	4,500
Garden and paths .	. . .	4,000
Fencing	. .	1,000
Water installation .	. . .	8,500
Sewerage	. . . .	2,000
Architect's fees	. . . .	3,000
Construction	.	73,714
Fitting up and sundry .	. .	22,286
Total	. . .	119,000 marks, or £5950,

making £170 per bed, or £158 without the ground.

[1] Kuthy, loc. cit. ; First Ann. Rep.   [2] Ibid.
[3] Rev. de la Tub., April, 1896, quoted from Belouet, Rev. d'Hyg., 20th
March, 1897.

Another estimate quoted by the same author is for the sanatorium in the Spessart for Würzburg, with 25 beds :—

	Marks.
Land and various installations .	10,000
Architect's fees and sundries .	15,000
Construction	75,000
Total . . . 100,000 marks, or £5000.	

This is at the rate of £200 per bed.

Beaulavon[1] gives the following figures :—

Ruppertshain, 3466 marks per bed.			Hanseatic Soc. 4140 marks per bed..		
Munich .	4460	,,	,,	Halle . . 5500 ,, ,,	
Albertsberg .	2515	,,	,,	Altona . . 2500 ,, ,,	
Brunswick .	8100	,,	,,		

The projected sanatorium for Nuremberg in the Engelthal is expected to cost 280,000 mks. for 60 beds.[2]

According to Liebe[3] the estimate for the sanatorium which is being built at Marcell in the Black Forest for 108 male patients amounts to £33,000, without electric lighting, and an additional £4200 for furniture. This is at the rate of £306 per bed without, or £344 with furniture. The recently-erected sanatorium at Altena in Westphalia was estimated to cost £15,000, with £1500 in addition for the doctor's house, and £150 for the chapel.[4] In Switzerland the sanatorium at Heiligenschwendi for 48 patients (now 52) cost £11,200, or £233 per bed ;[5] the Basel Sanatorium at Davos, without the ground, has cost about £13,500 for 61 beds (which will eventually be increased to 70), or at the rate of about £220 per bed.[6]

Coming next to the *cost of maintenance*, this is stated at Hohenhonnef to be from 11·7 to 13·72 mks. per head per diem,[7] but it must be considerably less at most of the other sanatoria for paying patients. For popular sanatoria, v. Leyden reckons the daily expenses in Germany at

---

[1] *Rev. de la Tub.*, Dec., 1896.　　　[2] *Heilst. Corr.*, July, 1898.
[3] *Hyg. Rundschau*, 1897, No. 21.　　　[4] Liebe, *loc. cit.*
[5] Kuthy, *loc. cit.*　　[6] First Annual Report.　　[7] Report for 1897.

2·50 mks.[1]   At Ruppertshain the daily cost is 2·77 mks.,
of which 1·22 is for food.[2]   It was 3·30 and 1·53 mks.
respectively at the old house at Falkenstein.   Ruppertshain
and Falkenstein are in the most expensive part of Germany.
At Malchow the daily cost is 3·16 mks.[2]   This is, however,
more a home for consumptives than a true sanatorium.   At
Grabowsee the cost per head is 2·47 mks. ;[3] at Rehburg
(Bremen Sanatorium), 2·26 mks. ;[2] at Blankenfelde, 3·35
mks. ;[2] at Dannenfels, 4·15 mks., or for food alone, 2·20
mks.[2]   The estimated daily cost at Marzell is 2·38 mks. ;[2]
at Würzburg Sanatorium, 2·19 mks. ;[3] at Altena, 3·02
mks.[2]   Both at Malchow and at Blankenfelde much
alcohol is used.   At Dannenfels there are only 18 beds.
The daily cost at the Königsberg Sanatorium in its first
year (1895) amounted to 2·31 mks., or with interest on
capital, 3·12 mks.[4]   In 1896 the cost was 2·23 mks., or
with interest, 2·71 mks.[2]   This sanatorium was formed
out of a pre-existing villa.

In Switzerland, the daily cost at the Heiligenschwendi
Sanatorium in its first (and most expensive) year amounted
to 1·92 frs.   At the Basel Sanatorium at Davos in its
first year the cost was as high as 4·09 frs., including
1·59 frs. for expenses of administration at Basel ; the
daily cost for food, drugs, washing and a few sundries
amounting to 2·2 frs.[5]

In France, Beaulavon[6] states that the average cost of a
consumptive in a general hospital is 2·93 frs. per diem ;
about one-third of the beds in the Paris hospitals are occupied by consumptives, whose average stay is ninety days or
more.   He estimates that the average cost per head in
these hospitals amounts to 3·15 frs., while in a sanatorium the daily cost would probably be only six centimes
more (3·21 frs.), and declares that whereas practically
no consumptive is ever cured in hospital, 28 per cent. are
more or less completely cured in a sanatorium.

[1] *Loc. cit.*   [2] Liebe, *loc. cit.*   [3] Kuthy, *loc. cit.*
[4] Annual Report for 1895-6.   [5] Annual Report for 1897.   [6] *Loc. cit.*

In London, the cost per head of in-patients in chest hospitals varies from 4s. 9d. to 6s. 6d. per day; at the Ventnor Hospital, which is practically a sanatorium, it is 5s. 4d.; so that we may expect the cost of maintenance in a country sanatorium in England to be no greater, and possibly less, than in a London chest hospital.[1] At the Adirondack Cottage Sanitarium in America, each patient costs a dollar a day. The average cost per day in most of the London general hospitals is considerably less than at the special hospitals, owing to the larger proportion of patients on low diet. Excepting for the absence of resident medical officers and nurses in convalescent homes, these would give a better standard for the probable expenditure in an English sanatorium for consumptives than a general or a chest hospital. Some of our convalescent and nursing homes receive consumptives; a few almost exclusively. But in order to draw trustworthy conclusions from the published reports, the average proportion of different stages admitted and the treatment adopted, as well as other circumstances, would have to be taken into account. The following estimate of probable expenses at the Altena Sanatorium (100 beds) is of interest :—[2]

	Marks.
34,400 days of treatment at 1·60 mks.	55,040
Physician, including dwelling	8,500
Assistant physician	2,000
3 sisters, 1800; 2 nurses, gardener, fireman, coachman, 2680	4,480
1 cook, 2 kitchenmaids, 2 housemaids	1,080
Food of staff: 12 at 1 mk.; 2 at 2 mks. . . .	5,840
Stocktaking, 2000; repairs, 1000; washing and soap, 1620	4,620
Water supply, heating and lighting	2,500
Bureau, 1000; horse, 1500; sundry and unforeseen .	2,890
Interest on capital, $3\frac{1}{4}$ °/$_o$; sinking fund, 1 °/$_o$ $= 4\frac{1}{4}$ °/$_o$ on 300,000 mks.	12,750
Total . . .	99,700

34,400 days at 3 mks. would yield 103,200 mks.

The expenses of maintenance of a sanatorium are partly dependent upon the size of the building, as a small estab-

---

[1] *Burdett's Hospitals and Charities*, 1897.   [2] Liebe, *loc. cit.*

lishment is relatively much more expensive. Those for women alone are a trifle less costly to maintain ; those for one sex alone a little less than those for both. Expenses may be considerably diminished by good arrangements and labour-saving contrivances, by the presence of water power, and of a railway at a reasonable distance; by efficient management, by work done by patients, and by the provision of a good orchard and kitchen garden, a fowl-house, pigsty and possibly a dairy farm. Fruit, vegetables, butter, eggs, milk and cream are indispensable in a sanatorium, and at certain seasons may be difficult to obtain in country places at reasonable cost, unless specially provided in the sanatorium.

The *charges* at sanatoria for paying patients vary between 36 and 91 mks. per week in Germany, in addition to an entrance fee and various extras. In France the prevailing charges are 14 to 20 frs. per day. In America it is difficult to draw the line between sanatoria for the less wealthy and those for the richer classes, as there is an almost unbroken series from those with a weekly charge of $5 to those with a charge of $20-30 or more.

In sanatoria for poorer patients in Germany the charges vary from $1\frac{1}{2}$ to 5 mks. per diem. At Arlen patients pay $1\frac{1}{4}$ to $2\frac{1}{2}$ mks. per day ; at Malchow and the recently-opened Felixstift at St. Andreasberg 2 mks. ; at Rehburg 2 mks. if from the Bremen Insurance Co., 3 mks. if from other parts of Germany, a few patients being occasionally admitted free of charge ; at Albertsberg they pay $2\frac{1}{2}$ mks., but a few free beds also exist ; at Grabowsee, Loslau, the Johanniter Isolir-Anstalt in Altena, and in Schömberg Sanatorium, the charge is 3 mks., with a few free beds at Schömberg. At Ruppertshain patients pay 3 mks. in the common rooms, or 5 mks. in the single-bedded rooms. A similar arrangement is in force at the new sanatorium at Altena, where $3\frac{1}{2}$ mks. is charged in the common rooms, and 5 mks. for a single-bedded room. The drawback to this arrangement is that the single-bedded rooms are likely to be frequently occupied by patients who do not medically need isolation as much as others, and who expect to be differently treated in other respects because of their extra payment. Moreover, it may be difficult to find accommodation for patients who are suddenly taken ill with some complication. There are three institutions which occupy an intermediate position between those for the poorer and those for the wealthier classes : the Brehmer second-class Sanatorium, the Krankenheim of Dr. Weicker (both at Görbersdorf), and Dr. Achtermann's second-class section at Laubbach. In Dr. Weicker's establishment the charges are 26 to 28 mks. per week ; in the other two 33 to 40 mks. with an entrance fee of 12 or 15 mks.

In America the Massachusetts State Hospital and the Montefiore Country Sanitarium are purely charitable institutions, as well as others which are hospitals rather than sanatoria. The De Peyster Children's Sanitarium charges $2½ per week ; the Adirondack Cottage Sanitarium and the Sharon Sanitarium, $5 per week ; at the Loomis Sanitarium the charge is $10-15 ; at White Gables Sanitarium, $6-22½ ; at the Las Vegas Sanitarium, $16 upwards ; at the Colorado Sanitarium, $13-35, with some free beds ; at the Hygeia, $15-30 ; at the Winyah Sanitarium, $20-30 ; at the Glockner Sanitarium, $8-15, with extra for medical attendance ; at the Home, Denver, $9, with medical attendance extra. In most of the above-mentioned establishments an extra charge is made for a special nurse, if required. At the Home, Denver, patients in need of constant nursing go into a special block, and pay $16 extra per week for the nursing.

In the Swiss sanatoria for paying patients the charges are about the same as in Germany. At Arosa the patients pay 9½ to 15 frs. per diem ; at Dr. Turban's Sanatorium at Davos, 13 to 17 frs. ; at Dr. Philippi's, 12 to 18 frs. ; at Leysin, 9 to 15 frs., exclusive of medical attendance ; at Montana, 9 to 15 frs., including medical fees. In each case an entrance fee is charged, in addition to various extras. Of the Swiss sanatoria for poorer patients, the charges at Heili-

genschwendi are from 1½ to 4 frs. per diem, according to
means ; at the Basel Sanatorium at Davos 5 frs., or under
certain circumstances 2 or 3 frs. ; at the Dutch Sanatorium in the same place, 4 frs. ; in the British Davos
Invalids' Home, 4 to 4½ frs. ; at St. Moritz, about 6 frs.
with the help of the "Aid Fund". It may be remembered
that at the Ventnor Consumption Hospital a charge is made
of 10s. 6d. per week.

Most of the above-mentioned charitable and semi-charitable institutions are more or less supported by private
subscriptions and donations. In Germany, however, there
is a law enforcing compulsory insurance against sickness
and old age, which has largely contributed to the foundation as well as the maintenance of popular sanatoria (see
p. 204). In Switzerland, where no such law exists, the first
sanatorium was founded by public subscription to commemorate a national anniversary ; and patients recommended by subscribers have the preference in admission.
In the Basel Sanatorium at Davos those who pay the full
amount (5 frs.) have first choice of the single-bedded
rooms ; certain societies and hospitals who contribute towards the funds have the privilege of sending patients (if
suitable) 'at reduced rates ; and in addition, an auxiliary
society exists to help those who are too poor to pay the
usual charges. At Davos, too, a society exists which secretly
assists those who are unable to continue paying the fees at
one of the pay sanatoria, on recommendation of the medical
officer of the institution.

In France the sanatoria for consumptive children and
youths at Ormesson and Villiers were established by private
subscriptions, being the first charitable institutions in
France to be started in this way. The Assistance Publique
and the Municipality of Paris, however, now subscribe to
the funds in return for certain privileges. The sanatoria
are free and non-sectarian, and open to the natives of every
country if resident in France. To stimulate the generosity
of donors to the children's sanatoria, certain honorary titles.

are bestowed according to the amount contributed ; and if
a sufficient sum be given, the names of the donors are in-
scribed on the wall of the vestibule at Villiers. Léon Petit
has suggested in his book[1] another expedient for raising
the necessary funds for popular sanatoria in France, an
expedient which in fact has been adopted in theory by the
Société des Sanatoria de France. It is that the rich shall
pay for the poor. In many German and Swiss sanatoria
for paying patients voluntary contributions are made towards the funds of the popular sanatoria ; but in Léon
Petit's plan this would be done systematically, by devoting
a portion of the profits of the pay sanatorium to the payment of the expenses at the popular sanatorium. He bases
his calculations on the figures furnished by Falkenstein and
Davos, and asserts that a sanatorium for one hundred paying patients would support another for fifty poor patients,
allowing for the payment of 3 per cent. on the joint capital.
Translated into pounds sterling, his table appears as follows :—

A.—Prime Cost.

1. Sanatorium for paying patients		£40,000
2. ,, poor ,, ...		10,000
		£50,000

B.—Annual Cost of Maintenance.

1. Sanatorium for paying patients—		
Daily cost at 6s. 4d.	£11,680	
Unforeseen	320	
		£12,000
2. Sanatorium for the poor—		
Daily cost at 4s.	£3,450	
Unforeseen	550	
		4,000
3. Interest on capital at 3 per cent.		1,500
4. Sinking fund at 5 per cent. (depreciation) . . .		2,500
		£20,000

[1] *Loc. cit.*, p. 258.

6

C.—RECEIPTS.

Sanatorium for paying patients—

Daily charges, 12s.	£21,880	
Sundries at 1s. 6d.	2,920	
		£24,800

I.e., profit nearly 10 per cent.

There is little doubt that were such a scheme started in England under proper auspices it would prove to be financially sound as well as socially beneficial. Owing to the greater cost of labour and material in this country, and the greater expense of living, it would not be safe to reckon on so large a dividend; but even if no profits were made, and the two institutions were together to be rendered self-supporting, they would be the means of effecting a great saving of useful lives and a corresponding diminution in poverty and suffering; while their indirect influence in educating the public in hygienic methods would be still more important.

CHAPTER XVII.

THERE are eleven or twelve sanatoria for the open-air treatment of consumptives in the Eastern States of America (one in process of erection), and seven more in the Western States, one of which, however, is not open for want of funds. Many of these admit other than consumptive patients; but the following profess to be exclusively for phthisical patients : The Adirondack Cottage Sanitarium, the Gabriels Sanitarium (being erected), the Loomis Sanitarium, the Sharon Sanitarium, the Winyah Sanitarium, and the Hygeia Sanitarium—all in the Eastern States. Of these, Winyah is a private institution for paying patients, the rest being charitable or semi-charitable. Most of the Western sanatoria are for paying patients. There are also a number of homes and hospitals for consumptives in the towns, intended mainly for more advanced and less hopeful cases. The term sanitarium is generally used in America in place of sanatorium. These two names might with advantage be used with stricter attention to their derivative meaning. The former might be applied to any specially healthy place ; the latter to a place with arrangements for restoring people to health— the former to an open health resort ; the latter to an establishment for hygienic treatment. The following is a list of the principal American sanatoria for consumptives :—

(83)

Adirondack Cottage San.	. New York State	1780 feet	94 beds
Gabriels ,,	. ,,	1970 ,,	50 ,,
Loomis ,,	. ,,	2300 ,,	90 ,,
Seaton ,,	. ,,	? ,,	? ,,
De Peyster ,,	. ,,	1100 ,,	40 ,,
Montefiore Country Home	. ,,	? ,,	40 ,,
Sharon San. . . .	. Massachusetts .	350 ,,	9 ,,
Mass. State Hosp. for Cons.	. ,,	1300 ,,	? ,,
Philad. Hosp. for Dis. Lungs .	Pennsylvania .	500 ,,	72 ,,
Winyah San. . . .	. N. Carolina .	2850 ,,	60 ,,
Aiken Cottages . . .	. S. Carolina .	? 350 ,,	? ,,
Hygeia San. . . .	. Alabama .	350 ,,	65 ,,
Glockner ,, . . .	. Colorado .	6000 ,,	40 ,,
Colorado ,, . . .	. ,,	5300 ,,	100 ,,
Bellevue ,, . . .	. ,,	? 5280 ,,	15 ,,
The Home, Denver . . .	,,	6250 ,,	22 ,,
White Gables San. . .	. Texas .	1428 ,,	25 ,,
St. Mary's ,, . .	. California .	? 4700 ,,	? ,,
Las Vegas ,, . .	. New Mexico .	6767 ,,	350 ,,

THE ADIRONDACK COTTAGE SANITARIUM

was founded in 1884 as an attempt to place within the
reach of working men and women the advantages of
climatic and sanatorium treatment at as moderate a cost
as possible.

It is situated one and a half miles from the village of
Saranac Lake, in the Adirondack Mountains, New York
State, on. sandy soil resting on primary rocks. The
grounds, which cover an area of about forty-five acres,
are surrounded on most sides by densely wooded heights
covered with evergreen trees, which protect the place
against wind. The sanatorium itself is several hundred
feet above the Saranac river valley, at an altitude of 1780
feet above the sea level, on a protected shelf-like plateau on
a hillside which slopes to the east and south-east. The prevailing winds are west and south. The climate in winter is
cold, with many windless snowy days. Snow lasts from
the middle of November to the middle of March or April.
Rain and snowstorms are frequent; and cloudy weather pre-
ponderates at all seasons, especially in winter. In February,
1894, the mean temperature was 13°, the minimum − 31° F.
In July, 1894, the mean was 66°, the maximum 91°. Gener-

FIG. 3.—THE ADIRONDACK COTTAGE SANITARIUM, NEW YORK STATE.
ADMINISTRATIVE BLOCK.

[*Face page* 85.

ally speaking, the climate may be characterised as cold and moist, but in winter cold and dry.[1]

The sanatorium was originally built to accommodate nine patients, but now has beds for ninety-four. It consists of twenty-two different buildings, from 75 to 100 feet apart, grouped around a large main structure on three floors. This contains the executive department, dining-room, offices, kitchen, baths, closets, nurses' and doctors' rooms, together with a few large rooms for a limited number of patients (fig. 3). The general sitting-room is 30 × 40 feet; the common dining-room, 40 × 50 feet. One of the buildings, enclosed by glass, is used as a recreation pavilion. Two of its sides (to the windward) are kept closed, the others being always open. Another building is used as an "infirmary," for the care of those who require to be kept in bed. The cottages are small one-storey buildings for from two to nine patients, the majority accommodating four or five. Each patient has his own bedroom, opening into a central sitting-room. The bedrooms average 10 × 14 feet, the sitting-rooms 18 × 25 feet. The partitions between the rooms are of solid masonry, but those between the bedrooms and the central sitting-room are only seven feet high, so as to give a larger ventilating space; moreover, the inner doors do not touch the floor. Ventilation is by means of open fireplaces in the central sitting-rooms, and by transoms over the verandahs and small openings over the windows of the bedrooms. The main building is ventilated by open fireplaces and ventilators in the ceiling, which lead by tin pipes to the chimneys. It is also heated by hot water. The cottages are heated by hot water, stoves and fireplaces. The whole sanatorium is lighted by electricity. The buildings are all constructed of hard wood and masonry, with as few angles as possible; the walls being of smooth varnished wood, without curtains or hangings, and the floors of hardwood, without carpets and with as few rugs as possible. Verandahs are placed outside the cottages, with glass par-

[1] Solly, *Medical Climatology*, London, 1897, pp. 210-214.

titions on the windy side. The water supply is from the
village waterworks, about two miles distant, and has a
pressure of 90 lb. The drinking water comes through
iron pipes from a spring 85 feet above the sanatorium, at a
point higher than any human habitation. The sewage is
discharged into the river 200 feet below the bluff on which
the sanatorium is built.

For disinfecting the rooms reliance is placed upon ample
ventilation, sunlight, and soap and water. The walls,
floors and furniture are washed daily with damp cloths.
Strict rules are enforced concerning the *sputa*. Each
patient is furnished daily with two small pasteboard spitcups in a frame. These are placed in a receptacle and
burnt in a crematory. In the halls, passages, and verandahs about four feet from the ground, are small boxes with
lids containing pasteboard glazed spitcups, which are burnt
and replaced every day. At meal-times Japanese paper
napkins are freely supplied and afterwards burnt.

The exercise taken by patients is prescribed by the
physicians. Those who show any appreciable rise of
temperature are only allowed to walk to the dining-room
for meals, and kept sitting out-of-doors during the febrile
period in the afternoon. If the temperature rises above
100° F. no exercise is allowed. The exercise taken by non-
febrile patients is determined by the state of circulation,
nutrition and appetite. Hydrotherapy is not used, but
patients have free access to the baths in the institution.
Few medicines are given, and principally those of a
reconstructive kind, such as cod-liver oil, hypophosphites,
arsenic, and, where indicated, creosote in small doses. No
other antiseptics or specific remedies are given, excepting in
a few selected cases which have been treated with tuberculin.

The sanatorium is under private management, State aid
having been refused. It is directed by its founder, Dr.
E. L. Trudeau, and a board of trustees, with the help of
two resident physicians, examining physicians in New York
and other large cities, and trained nurses in the "infirmary".

The charges are $5 per week to all alike, the deficiency (about $2) being covered by subscriptions and contributions. There are no extras, excepting for extra washing, medicines (supplied at cost price), and a small charge for nursing in the " infirmary ".

The average length of stay is six months ; one year is the limit ; but in exceptional cases when necessary this rule is not enforced. Patients who show no improvement after a reasonable period are advised to return home. Advanced cases and active types of disease are only exceptionally admitted. Those who can afford to pay more than the low prices charged are not admitted. Friends are only allowed to remain a week, and pay $1·50 per day. The institution is non-sectarian ; there is a stone church in the grounds, in which services are held nearly every Sunday by clergymen of any denomination who may volunteer their services.

Statistics.—20 to 25 per cent. are apparently cured, and in 20 to 35 per cent. more the disease is more or less permanently arrested (mean results of ten years). If the most favourable of the cases admitted are separated under the term " incipient," the proportion of patients cured is as high as from 30 to 35 per cent.

Saranac Lake village can be reached in thirteen hours by rail from New York without change of carriage. The patient can have all his meals brought to him in the train.

THE GABRIELS SANITARIUM

is being erected in the Adirondack Mountains, at an altitude of about 1970 ft., near Paul Smith's, Franklin Co., New York, and will have fifty beds, under the care of Catholic Sisters. Paul Smith's is on the north shore of the Lower St. Regis Lake, a lovely chain of lakes about five miles long, with sandy shores and very little rock. The neighbourhood is comparatively level, with one mountain (St. Regis), about 3000 ft. high.[1]

[1] S. E. Solly, *loc. cit.*, p. 210.

THE LOOMIS SANITARIUM

is situated about 120 miles from New York, at Liberty, Sullivan Co., in the midst of a rolling mountainous country, which forms the highest land between New York and the great lakes. The Catskill Mountains are some twenty miles to the north and east; southward is a wide valley for forty miles; westwards the view extends over the tops of hills for fifty miles along the river Delaware into Pennsylvania. The value of this beautiful country was advocated years ago by Dr. St. John Roosa and the late Dr. Alfred Loomis. There are practically only two seasons, winter and summer, with a rapid transition between the two. The winters are cold and dry, while even in summer the air is dry and exhilarating and the nights cool and refreshing.

The sanatorium was erected by public subscription and opened on 1st June, 1896, with accommodation for twelve patients, but has now beds for ninety with additional tent accommodation in summer. It stands 2300 feet above the sea level on a southerly slope, protected by a rocky wooded ridge to the north. The grounds belonging to the institution have an extent of 193 acres, and form part of a very large tract of open country. The grounds have been laid out with roads, paths and terraces.

The sanatorium consists of a group of buildings scattered across the turf and connected with paths. The largest is the administration building (fig. 4), which is a picturesque three-storey building with verandahs and round pointed roofs, and out-buildings attached in a long line beside it. This contains on the ground floor the reception-room, library, dining-room, offices, drug-room, butler's pantry, kitchen, store-room and laundry. On the next floor are the solarium, four emergency wards, laboratory, nurses' rooms, baths and closets, guest rooms and sleeping quarters for the resident staff. On the third floor are servants' quarters and store rooms. Another building, the casino, is a two-

FIG. 4.—THE LOOMIS SANITARIUM, NEW YORK STATE. ADMINISTRATIVE BUILDING.

[*Face page* 88.

Fig. 5.—The Loomis Sanitarium, New York State. Marcy Lester Cottage. [Face page 89.

storey building with pointed roof and large open verandah
devoted to amusement, and containing a billiard table,
piano, harmonium, etc. Two other cottages of two storeys
each accommodate respectively twenty-three and sixteen
patients. Five other one-storey cottages accommodate
eight, eight, four, four and five patients respectively.
There are two other buildings, of which one accommodates
seven, the other ten patients. The buildings are mostly flat
on one side and rounded on the other with verandahs which
are protected in winter by glass at the exposed ends, but
remain open to the south (fig. 5). There are bathrooms and
closets in every cottage, averaging one to every four patients.
Each cottage has one or two parlours. The bedrooms average twelve by fourteen feet in size, some provided with
chimneys. Nearly all accommodate but one patient, eight
have beds for two a piece. Each building is heated by its
own hot water or steam pipe plant, and is lighted with
electricity from a central dynamo. Ventilation is by open
windows, through the steam heating level, and through
the open fireplaces. The temperature is not allowed to
exceed 65° F. by day and 40° by night in the winter
months.

The walls and floors are capable of easy cleansing and
disinfection. The furnishing is comfortable but simple,
without unnecessary hangings, and without fixed carpets.
The water supply is from two large springs which fill a
reservoir above the highest part of the sanatorium. The
sewage is disposed of according to the Waring system.
Trained nurses are kept in the sanatorium. There is also
a nurses' training school consisting at present of thirteen
members who are given practical training every day, and
lectures and recitations every four days. The superintendent is a graduate of a New York training school; the
course of training of two years' duration. The physician
in charge (Dr. J. E. Stubbert) lives in a separate house; a
house physician and assistant house physician are lodged
in the administration building. Service is held every

Sunday in the same building. A memorial chapel will probably soon be built.

The treatment consists partly of hygienic, partly of other recognised methods. There are no general assembly rooms, beyond the dining-room and casino, as patients are expected,. and are able as a rule, to spend nearly all their time out of doors. Most of them take a large amount of exercise. Some cycle, play tennis, or ride. Breathing exercises are taught by the house physicians. There is no special provision for exercise in wet weather. Three regular *meals* a day are provided in the general dining-room. Hydropathy, counter-- irritants and cod-liver oil are little used. Antistreptococ- cus serum, ichthyol, creosote, oil of cinnamon, and oil of cloves are administered and stated to give good results in the order named. Hot-air antiseptic inhalations and local treatment for laryngeal tuberculosis are employed when necessary, and occasionally Koch's tuberculin and similar remedies. Only early cases are admitted, which are ex-- pected to benefit by the treatment and to offer a chance of cure. A careful record is kept of every case, *sputa* being examined for bacilli, and the Röntgen rays being used to confirm the results of auscultation and percussion. In the case of sudden and serious illness, patients are transferred to the infirmary. The usual duration of treatment is about six months.

From 1st June, 1896, to 31st March, 1897, eighty-seven patients were admitted into the sanatorium, forty-six being men, forty-one women. Out of these, forty-three were discharged ; forty-four remaining under treatment. The results in those who were discharged were as follows :—

CLASS I.—REMAINING THREE MONTHS OR LESS.

On Admission.		*When Discharged.*	
Incipient stage, without bacilli	3	Apparently cured . . .	4
„ „ with „	4	Disease arrested . . .	2
Moderately advanced .	9	Improved	12
Far advanced . . .	13	Unimproved . . .	11
	29		29.

Class II.—Remaining More Than Three Months.

On Admission.		When Discharged.	
Incipient stage, without bacilli	3	Apparently cured . . .	3
„ „ with „	2	Disease arrested . .	1
Moderately advanced . ".	6	Improved	7
Far advanced	3	Unimproved	3
	14		14

So that 67·5 per cent. of those treated showed improvement, and in over 23 per cent. the disease was arrested or apparently cured.

The fees charged are $10-15 per week, or $7 where two occupy the same bedroom. This includes medical attendance. Medicines, laundry and additional nurses, when required, are extra. Friends pay $2 per day. These charges are only sufficient to render the sanatorium self-supporting. The medical management is vested in a medical board which is mainly in New York, but is represented at the sanatorium by the physician in charge. Business matters are managed by a board of ladies in New York, and at the sanatorium by a lady superintendent.

Liberty is a station on the main line of the New York, Ontario and Western Railway, and is about three and a half hours by rail from New York. The sanatorium is on the telephone.

The De Peyster Hospital

is situated a few miles from Verbank Station in Dutchess Co., New York, about seventy-five miles from the capital, and is intended for consumptive children of the poor between two and sixteen years of age. It stands about 1100 feet above the sea-level, with a southern exposure, and protected to the north by pine woods. It is heated by steam, and has a long covered balcony on the southern side.

The hospital was started this year, and is unendowed, being supported by voluntary subscriptions, and under deaconess' management. Only curable cases are admitted, at a charge of $2.50 per week, including medical attendance and drugs. There is accommodation for forty children.

THE MONTEFIORE HOME COUNTRY SANITARIUM

is under the same management as the Montefiore Home for Chronic Invalids at New York, described at p. 117.

It is situated at Bedford Station, Westchester Co., New York State, in a sheltered situation on the Berkshire Hills, which form the highest part of the county, and is sixty miles from New York or one and a half hours by rail.

Opened early in 1897 with beds for ten patients, it was very soon enlarged to accommodate forty ; and it is hoped to further extend the number of beds to sixty.

The two buildings are frame houses, one of which has a large verandah. There is also a good bathroom, and heating by hot water.

The grounds cover 136 acres of land. It was originally a farm ; and patients are sent there who are able to do a little light work, with the object of ultimately making the sanatorium self - supporting. It has already begun to supply the Home in New York with milk and eggs. Nearly all the patients are consumptives in an early stage ; but a few are sufferers from asthma, neurasthenia, etc. Only men are admitted, and no charge of any kind is made. There is at present no house physician ; but one of the visiting physicians of the Home in New York attends once a week, and more frequently if called through the telephone.

THE SEATON SANITARIUM

is situated at Spuyten Duyvil, N.Y., and has two visiting physicians, Dr. Jackson and Dr. Shrady, but no house physician. I have, however, no further information about it.

SHARON SANITARIUM,

which was founded by Dr. V. Y. Bowditch in 1890 by means of public subscription, is situated about one and a half miles from the village of Sharon, five minutes' walk from the

station on the Providence Railway, and eighteen miles from
Boston. It stands amidst hilly country covered with pine
woods, on thirty feet of gravel and sand, at a height of 300
to 400 feet above the sea. The climate is moist and variable.
The sanatorium is on the southern slope of a hill, in about
130 to 140 acres of woodland, protected by hills to the N.
and W., and by thick pine woods in the immediate neighbourhood, and has a large amount of open land to the south.
It is a large wooden building with accommodation for nine
patients, and has been constructed so as to obtain as much
fresh air and sunlight as possible by means of numerous
windows and open fireplaces in every room. Each patient
has her own special bedroom, which averages 10 × 12 feet.
The walls are painted, the floors of hardwood, the corners
rounded in the newer parts. Rugs are used instead of
fixed carpets, and no dusting or sweeping is done, the rooms
being cleansed with damp cloths, which are afterwards
boiled or burnt. The heating is by a hot-air furnace, with
hot-water attachment in the bathrooms. There is a good
supply of pure spring water. The sewage runs into a cesspool with an overflow 200 yards from the house. There
are broad piazzas for rest in the open air, part covered with
a verandah.

The place is intended for consumptive women only of the
refined but not wealthy classes—such as school teachers,
needlewomen and the like. Only cases in an early stage
are admitted. It is managed by a matron who is a trained
nurse, other nurses being obtained if necessary. The medical director is Dr. V. Y. Bowditch, of Boston, whose assistant visits the sanatorium every day. The treatment is
for the most part hygienic—good food, fresh air, regular
hours, and carefully regulated exercise. Three *meals* per
day are provided, with two lunches of milk and eggs. The
windows are always kept open ; febrile patients rest in
bed ; others rest on the piazza in all weathers, or take
increasing exercise according to their condition, always
stopping short of fatigue. Respiratory exercises and the

pneumatic cabinet are used in all suitable cases. Hydropathy is not employed excepting in the shape of cold sponging. Cod-liver oil and counter-irritants are occasionally
used. Klebs' antiphthisin has also been tried in a few cases.
Hypophosphites, bitters, malt extract, and iron are the drugs
chiefly used ; and special attention is also paid to the state
of the stomach and bowels. The *sputa* are received into
" sanitas paper sputa cups," and afterwards burnt. Patients
in the grounds carry rubber pouches with Japanese paper
napkins, which are also used at meal times and destroyed
by burning. Linen is disinfected by boiling. The rooms
after a patient leaves are disinfected with chloride of lime
solution. The charges are $5 per week, including all but
washing.

The results are very satisfactory, when it is remembered
that the sanatorium possesses no special advantages from
climate or elevation above the sea level, and is situated in a
district where phthisis abounds. It may fairly be said that
Dr. Bowditch has proved that the treatment can be satisfactorily carried out within a very moderate distance of a
large town or city, and with very simple means. He stated
in June, 1896, that out of sixty-four cases treated during
the last five years the disease was arrested in over one-third
(twenty-two cases). In all but six the arrest continued, and
the patients had remained in good health, mostly for more
than two years. One had died after an operation for uterine
disease without recurrence of the pulmonary symptoms ;
three others had died of phthisis ; but two of these had
resumed an unhealthy mode of life contrary to advice.

THE MASSACHUSETTS STATE HOSPITAL FOR CONSUMPTION
has recently been finished (January, 1898), and is to
be immediately opened. It is situated in Worcester Co.,
near Rutland, 1300 feet above the sea. It consists of a
number of one-storey wooden buildings of various sizes,
arranged in a semicircle, with the administrative block in
the centre. On one side are pavilions for men, on the

other for women. The larger buildings have each seven private rooms and one large ward for twenty-two patients. The smaller have only a ward for ten patients. Each pavilion has its own solarium constructed entirely of glass, and is surrounded by a broad verandah.[1]

THE PHILADELPHIA HOSPITAL FOR DISEASES OF THE LUNGS

belongs to the Protestant Episcopal Mission of that city, and is managed by a board appointed by the bishop of the diocese. It includes a small temporary hospital for men (House of Mercy) in the heart of the city, and a larger hospital for women at Chestnut Hill. These institutions are mainly, but not exclusively, intended for the reception of consumptives, and formerly went under the name of Homes for Consumptives. The patients are for the most part poor people, who are admitted free of charge ; but well-to-do patients also enter at times, and make a dona-tion to the funds, there being no stated charges for such cases. All stages are admitted, the length of stay being usually three months, prolonged to six, or more if advis-able. The mission also has a number of sick-diet kitchens in the city, and helps the poor in this way in their own homes. Last year (1897) 164 were admitted to the two hospitals, and thirty-six helped in their own homes with food and money. From 25 to 33 per cent. of those ad-mitted to the hospitals are said to be apparently cured, a fair proportion permanently. The medical staff consists of two resident officers, a clinical clerk, and eight visiting physicians. There are altogether nine nurses.

THE HOUSE OF MERCY,

or male branch, is at 411 Spruce Street, next to the general offices of the mission. It consists of an ordinary house, which has been adapted for the purpose, and has twelve beds. This will probably be replaced by a more permanent and more suitable building.

[1] *Heilstätten Correspondenz*, Jan., 1898.

THE CHESTNUT HILL HOSPITAL

is within twelve miles of the City Hall, in a residential
suburb of Philadelphia at its most northerly part. It
is surrounded by gentlemen's country seats of one to two
acres, and has farm lands to the north, being practically in
the country. The mean temperature of this part is some
6° colder than that of Philadelphia itself. The soil is
of limestone. There are about thirteen acres of ground
belonging to the hospital, partly wooded and partly open.
The sanatorium is about 500 ft. above the sea, on the crest
of a hill which forms the highest ground for miles. There
is no special shelter against wind from hills or trees. The
buildings are on the southern slope, and consist of a central
administrative block, with a number of cottages on each
side, mostly built within the last ten years and accommodating sixty patients. The administrative building
contains the day rooms, doctors' quarters, dining-room,
kitchen and emergency room. There is also a small
kitchen in each cottage. The patients' bedrooms are
all to the south, corridors with double windows to the
north. The rooms are large, and mostly contain but one
bed, a very few having two. Most of them have chimneys.
The walls are undecorated, the corners rounded. Every
cottage has a large common room, which is used as a sun
parlour and for other purposes. The heating is by hot air;
fresh air entering an empty space under the building, where
it is heated by radiation from a drum, and then driven into
the rooms. The lighting is by the city gas supply; the
sewerage and water supply being also connected with the
city systems. There is ample provision for both rest and
exercise in the open air without exposure to bad weather.
Patients rest either in their own rooms, with windows
open, or in the sun parlours, winter as well as summer.
The douche is not used. Very little cod-liver oil is given.
At the House of Mercy the *sputa* are received into porcelain
spitcups, which are disinfected with corrosive sublimate

solution ; at Chestnut Hill paper spitcups are used, which are afterwards burnt. Linen is disinfected by boiling. Rooms were formerly disinfected by 10 per cent. carbolic solution and by burning large quantities of sulphur. Now formaldehyde vapour is used instead. I am indebted for these particulars to Dr. William M. Angney, consulting physician.

THE WINYAH SANITARIUM,

at Asheville, North Carolina, was founded in 1878 by Dr. J. W. Ghitsman, of New York City ; remodelled in 1888.

It is situated on lime and sandstone substrata on a south-westerly slope at a height of 2350 feet above the sea, with mountains to the north which shelter it from prevailing winds.

The grounds are about four acres, and touch the woods on the mountain slope.

The sanatorium is a three-storey brick building with accommodation for sixty patients, and has toilet and bath rooms on each floor. It contains three general assembly rooms, measuring about 40 × 60 feet. The bedrooms average 16 × 16 feet, and are provided with chimneys. Ventilation is by air shafts and open fireplaces. In addition to these, the building is heated by steam pipes. The lighting is by electricity. The water supply and sewerage are those of the city of Asheville. The establishment is under the direction and control of Dr. Karl v. Ruck, to whom it belongs. There are also an assistant physician and a resident house physician. There are no trained nurses, excepting as called in for emergencies.

Pay patients alone are admitted, and no hopelessly advanced cases are received. Patients take their meals in the general dining-room or, if they cough and expectorate freely, in their own rooms or on the piazza surrounding the building. They take exercise under daily medical prescription, according to the state of the heart and the body temperature. The cold rub is administered in all

7

cases, the cold douche in selected cases. Counter-irritants
are not often used. Tuberculin is made use of in special
cases. *Sputa* are received into spitting-flasks, and afterwards destroyed by fire. Handkerchiefs are permitted,
but not for expectorating. Linen and eating and drinking
utensils are purified by steam or boiling water. Rooms
are disinfected by the generation of formaldehyde. They
are constructed so as to facilitate cleansing.

The charges are $20 to $30 per week ; no extras.

Statistics : Early stage . apparent recovery 81 %, improvement 19 %
More advanced .   ,,   ,,   35 %,   ,,   23 %
Advanced,   but⎫
  still in fair ⎬ ,,   ,,   9 %,   ,,   24 %
  condition   .⎭

The average length of stay is six months.

The sanatorium is easily accessible from Asheville by
electric tramcars.

THE AIKEN COTTAGES,

at Aiken, South Carolina, have been mentioned to me by Dr.
Trudeau as a sanatorium for consumptives. Aiken stands
550 feet above the sea-level on an elevated plateau between
the Savannah and Edisto rivers, a little over a hundred
miles from the ocean. It lies on sandy soil amidst pine
woods. The climate is mild ; there is a considerable rainfall (average 48 in.), but not much cloudy weather in
winter, when the relative humidity is 59 per cent.[1] There
is said to be a good water supply.

DR. SAJOUS' CO-OPERATIVE VILLAGE.

In the *British Medical Journal* for 31st October, 1891,
it is stated that Dr. Sajous had started a company for
supplying middle class consumptives with model cottages
in suitable places under sanitary and medical supervision.
The scheme was to be co-operative ; provisions, trained
servants and nurses being supplied by the company. The

[1] S. E. Solly, *loc. cit.*, pp. 223-4.

rent of a furnished cottage was to vary from $20 to $60 per month, and the first of these villages was to be founded at New Florence, near the Gulf of Mexico. I have been unable to learn whether the scheme was ever carried out.

THE HYGEIA

is situated one mile from Citronelle, Alabama, a town of about 3000 inhabitants, free from factories, on the Mobile and Ohio railroad, and has a special station of its own. It is surrounded by rolling pine-clad country, and stands in 260 acres of land, part of which is laid out as a park, with walks and roads, croquet grounds, golf links and tennis court. Shooting and (at a distance) fishing can also be obtained in the neighbourhood.

The climate is sub-tropical, equable owing to the proximity of the Gulf Stream, yet free from extreme humidity and fogs, and mild in winter without being oppressive. There is usually very little rain, excepting for a short time in January.

The sanatorium, which is open from 15th October to 15th May, is on sandy soil, 350 feet above the sea-level, and is sheltered from winds by the extensive pine woods to north, east and west. It consists of three large buildings and five cottages, all of them wooden-frame houses with wide porches, and balconies or verandahs on every floor. The main building, which was built in 1889, consists of a dining-room, parlours and reading-room, the office and a certain number of bedrooms ; while an annexe contains the kitchen and pantry, bath-rooms and barber's shop. Another building contains a billiard-room and a bowling alley, also a drug store, newspapers, books, etc. There is a separate building for consulting rooms and inhalation room, with arrangements for inhalation of medicated vapours and compressed air. The bedrooms average 15 × 15 feet, some at a lower price being 12 × 8 feet ; all are to the south in the cottages, those in the main building having various aspects.

The walls are oil-painted or calcimined, the angles being partially rounded, and no prominent decorations being added. The heating is by open fireplaces; the lighting will probably be by electricity, but it is at present by oil. Ventilation is by means of open windows, transoms and other special appliances. The sewerage is carried into a stream one and a half miles off. The water supply is from springs and pumped into tanks.

The sanatorium, which is a private institution exclusively for consumptives of both sexes in presumably curable stages, has sixty-five beds. Only paying patients are received, but some are admitted at reduced rates into one of the buildings. The treatment is mainly hygienic, rest or exercise in the open air being regarded as of supreme importance. The douche is not used, but in most cases cold affusion with sponging and friction, a few receiving the dry rub instead. Cod-liver oil is not given excepting in some cases to children. Tuberculin, tuberculocidin, preventive serum and the new tuberculin T.R., have all been used in small doses in a certain number of cases, and are regarded as of value. Counter-irritation by iodine or faradisation is also occasionally employed. For the *sputa* the use of handkerchiefs is strictly prohibited, enamelled spittoons being used instead, with 1 in 1000 corrosive sublimate, the contents being afterwards sterilised by heat Linen is steam-sterilised; rooms disinfected by formalin vapour. At the most two patients are allowed to sleep in one room. The average length of treatment is about 130 days, the longest stay having been 240 days. Patients are not advised to leave before three months in any case.

There are two or three resident medical officers, the chief being Dr. A. C. Klebs; three consulting medical officers, including Dr. Edwin Klebs of Chicago. There are also a general manager and a staff of three nurses, others being obtained if necessary. The charges are from $15 to $30, including board, lodging, baths and ordinary medical and nursing attendance.

The results of the two winters, 1895-6 and 1896-7, were as follows :—

> On admission 31·25 per cent. were in first stage, 32·50 per cent. in second, 36·25 per cent. in third.
>
> In 22·50 per cent. all signs and symptoms disappeared.
>
> In 46·25 per cent. there was decided objective improvement.
>
> 16·25 per cent. remained stationary.
>
> 2·50 per cent. died.
>
> In 65 per cent. there was gain of weight.
>
> In 62·50 per cent. the fever diminished, in 32·50 per cent. it remained stationary, in 5 per cent. it increased.
>
> The bacilli were absent on last examination in 40 per cent.[1]

"The Necessity of Special Institutions for the Treatment of Pulmonary Tuberculosis," by A. C. Klebs, *Tri-State Med. Journ. and Pract.*, St. Louis, May, 1897. Also private letter.

CHAPTER XVIII.

SANATORIA IN AMERICA—WESTERN STATES.

THE GLOCKNER SANITARIUM,

at Colorado Springs, is situated at an altitude of 6000 feet above the sea-level on a plateau at the northern end of the city, and two miles from its centre, being easily accessible by electric cars. The surrounding country is mountainous to the west and north, with plains to the east and south. The soil consists of gravel and sand, seventy feet thick, lying on a clay bed which slopes to the south. The Rocky Mountains and foothills surround the city on all sides and protect it against wind. The sanatorium, which was specially built for invalids by Mrs. Glockner in memory of her husband, Mr. Albert Glockner, stands in a garden 400 feet square, and has accommodation for about forty patients, consumptives in all stages being admitted as well as other patients. It is managed by sisters of charity and has trained nurses but no regular medical staff, the patients calling in whatever doctor they please. The building is of brick, with a southerly aspect, and is so constructed that every room receives the sun, and every part is readily cleansed. It is said to have every modern convenience and appliance, with porches and covered balconies on every storey, some enclosed by glass. It is heated by steam, the assembly rooms and some of the bedrooms being provided with chimneys. Ventilation is by means of transoms and windows. The assembly rooms are 52 × 16 ft.: the bedrooms averaging 12 × 18 ft. The place is lighted by electricity. There is a good water supply. The sewerage

(102)

is connected with the city system. The *sputa* are received into sanitary cuspidores which are afterwards burnt. The rooms are disinfected by formaldehyde gas. The *meals* are : breakfast at 8 A.M., dinner at 1 P.M., supper at 6 P.M. ; milk, meat juice and eggs as requested between meals. Massage is often used ; counter-irritants and cod-liver oil in suitable cases, but no specifics as a rule. Patients take plenty of exercise if their strength permits. Hydropathy is not employed, excepting ordinary baths. The charges are from $8 to $15 per week for board and lodging ; medicines, stimulants and medical attendance being extra.

THE HOME, DENVER, COLORADO,

was founded in September, 1895, by the Council of the Episcopal Church of Colorado, with the help of subscriptions from wealthy people in New York and elsewhere. The object was to establish a home for consumptive Christian ladies and gentlemen of limited means, mainly of the professional classes. A reference is required as to character before admission. The sanatorium is chiefly intended for early and presumably curable cases, but every stage is received and provision is made for nursing those who require it.

The home is under the superintendence of the Rev. F. W. Oakes, with the help of a manager who has had long experience in both hotels and hospitals. There is a Board of Management consisting of the Rev. F. W. Oakes, Dr. Samuel A. Fisk and Rev. David H. Moffat, under the presidency of the Rt. Rev. Dr. Spalding. There is no medical resident, but each patient selects his own medical adviser from Denver or elsewhere.

The home is about ten minutes' ride by cab or tramcar from the post office at Denver on an elevation commanding an extensive view over the city, the Rocky Mountains for about 150 miles and the plains to the east for several hundred miles. An uninterrupted view may be had from the porches of Pike's Peak, Mount Evans and Long's Peak.

The sanatorium lies at an elevation of 6250 feet above the sea-level, on three or four acres of ground which slope to the south and east. The Rocky Mountains protect it to the west.

There are no trees within 500 feet of the building. The soil is very porous and sandy. The sanatorium consists of four separate buildings (fig. 6), united by a roomy, well-lighted corridor, which itself forms a sort of elongated sitting-room, and the roof of which is used as an open-air lounge. One building (" St. Andrew's House ") is for men ; another (" Grace House ") is for husband and wife, or mother and son ; next in order is the " Emily House " for women ; and next to this " Heartsease," for those in need of nursing. The three former consist of two floors and an attic floor. There are sitting-rooms in every house and covered porches front and back surmounted by large balconies with awnings; also a music room ; a library with 2000 volumes ; a gymnasium with billiard tables, dumb-bells, etc. ; a large dining saloon ; and an abundance of bathrooms, lavatories and closets. The place is most luxuriously furnished and decorated with valuable pictures, rich carpets, and all the comforts and conveniences of a well-appointed home. All walls are painted in oil, the angles being square. Floors are all varnished, the rooms are all sunny and of good size, none less than 14 × 15 ft. and $11\frac{1}{2}$ ft. high, some are 18 × 20 ft. Every room, including bedrooms, is furnished with a fireplace and chimney, and separated from the next by a brick wall. " Heartsease," which was built in 1897, is connected with the rest of the establishment by a covered corridor with tiled floor. It is a four-storey building. The basement, which is partly above ground, contains kitchen, pantry and servants' quarters ; the ground floor has a large and beautiful sitting-room with large bow windows and window seats ; a dining-room, diet kitchen, bath-rooms and seven large bedrooms ; the upper floors contain more patients' rooms, of which there are twenty-two in all. Every partition consists of a thick (15″) brick wall,

[Face page 104.

Fig. 6.—The Home, Denver.

which is practically sound-proof. There is only one bed in each room, the walls and furniture being of much the same character as in other parts of the establishment. There is a diet kitchen on each floor where light cooking is done by electricity. This kitchen is connected with the main kitchen by a dumb-waiter. Each room has its separate silver service, linen and china, and each floor is in charge of trained nurses, of whom there are one to every six patients, besides the superintendent nurse. Every floor has its own baths, etc., with tile, marble and porcelain fittings. There is a lift to every floor, and a well-stocked drug room. The whole house is in charge of a skilled and trained nurse. matron.

The home is heated by indirect central steam heating, and lighted by electricity. The water supply, which is cool, soft and abundant, is from a private artesian well. The house is connected with the city sewers, and its sanitary arrangements are said to be very good. Ventilation is by the fan system. There are two 100 in. fans in the basement forcing fresh air into every room and replacing the air twelve times per hour. Windows are kept open whenever it seems advisable. The *sputa* are exclusively received into cuspidores, which are cleansed four or five times a day with a strong solution of corrosive sublimate. The linen is steam-cleaned ; the rooms disinfected with formaldehyde and then washed ; bedding and rugs sterilised by heat. Most of the patients who enter the home are fairly well and able to take plenty of exercise. There is accommodation for altogether 100 visitors. Three *meals* a day are provided : at 7·30 to 9 A.M., 1 to 2 P.M., and 6 to 7 P.M.

The charges are $9 per week for board and room, or $25 in " Heartsease," also including the nursing, but not the medical fees. These charges, which only pay about half the expenses, are supplemented by subscriptions. There is one endowed room for a male visitor. A chapel is soon to be built, which will be lighted at night and always kept open.

· LAS VEGAS SANITARIUM

is situated at Las Vegas Hot Springs in the northern
portion of New Mexico, U.S.A.   Las Vegas Hot Springs is
a health resort six miles from Las Vegas, on a branch of the
Santa Fé Railway, which has five trains each way per
diem.   It is under the control of the railway company, and
forms a village entirely devoted to the needs of visitors and
invalids, and specially laid out with a view to hygienic
requirements.   The position and climate are unusually
good.   Standing in the same latitude as the African
Sahara, it is at a considerable elevation, so that hot weather
is unknown.   Blankets are required every night of the
year, and the summers are delightfully cool.   During the
two summers of 1896 and 1897 the temperature never rose
to 90°, and in 1897 it did not reach 80°.   In winter there
is sometimes snowy weather, but as the place shares in the
dryness characteristic of the Rocky Mountain region, the
cold is beneficial to tubercular conditions without being
disagreeable.   The average winter sun temperature is 76°.
The daily average is said to be 20° higher in winter than
at Denver, and 20° lower in summer.   The average annual
rainfall is 12·7″.   From September to June the average
is considerably less than 1″.   The average number of
sunny days is 326.   The sanatorium stands on granite
and sandstone, at a height of 6767 feet above the sea-
level, at the entrance of a cañon of the Rocky Mountains,
in the Canadian valley (fig. 7).   Neighbouring peaks of the
true Rocky Mountain range to the west rise to heights
of 10,000 and 14,000 feet; north and south are spurs from
the same range, and all round, forming four-fifths of a
circle, are the foothills.   There is consequently very complete protection against wind, and sandstorms are said not
to occur there.   Owing to the dryness of the air, it is quite
possible for invalids to remain out of doors day and night;
but the precipitation is just sufficient to ensure the growth
of vegetation in the neighbourhood.   Over 500 acres of

FIG. 7.—LAS VEGAS SANITARIUM.

[Face page 106.

FIG. 8.—LAS VEGAS SANITARIUM.

[Face page 107.

land belong to the place, through which runs the Gallinas river. There are about thirty hot springs in the place, ranging in temperature as high as 144° F., and varying in chemical constitution from saline to lithia, iron and sulphur springs. Some of these are conveyed by pipes to the sanatorium. The mountains around are covered with pine trees, cotton woods, cedars, and the like. The sanatorium consists of one large (fig. 8) and several smaller buildings grouped round a central park of fifty acres. The main building (the " Montezuma ") accommodates 250 visitors; another ("Mountain House ") will hold sixty ; and forty or fifty more can be lodged in cottages and at the bath establishment. There is a dining saloon to seat 250 people, and a lecture hall for 500. The main building is of stone, and consists of an irregular oblong mass, with a round tower at one end furnished with a spire and with a circular sun gallery just under the roof. At the foot of the building is a verandah 15 feet wide and 544 feet long, protected by 70 feet of glass in case of bad weather. The bedrooms average 13 × 15 and 13 feet high, but some are much larger. The whole establishment, including the cottages, is heated by steam pipes, and lighted by electricity. There are no stoves or chimneys. Pipes convey fresh air into each room, and independent flues carry out the foul air through the roof. The sewerage is according to the Bertin system. A reservoir 200 feet above the sanatorium hewn out of the solid rock and containing 4,000,000 gallons is filled with water pumped out of the stream, and flushes the sewers. These are carried down 150 feet lower and one mile away to a farm of fifty acres and purified by irrigation. The drinking water is abundant and entirely from natural springs. The *meals* are at 7·30 to 9·30, 1 to 2·30, 6 to 8 ; and an early breakfast at 6 to 7 A.M. in the bedroom if desired. Two lunches are served free out of doors at 11 A.M. and 4 P.M. The supplies come partly from the farm and from hothouses belonging to the establishment. A herd of goats is also kept. A bath-house 140 feet long provides swimming and various hydropathic appliances. In another house peat baths are provided. There is a hospital with all modern appliances facing the park some distance from the main building. Attached to this are trained nurses and a training school. Other buildings connected with the sanatorium are a casino, post office, station house, power house, school house, barn with stables, telegraph and telephone office, open day and night, as well as express and money order office. The sanatorium is officially recognised as a meteorological station. Altogether 1¼ million dollars have been spent upon the place to render it as perfect as possible for both invalids and pleasure-seekers. No advanced cases of phthisis are received, nor any infectious or otherwise obnoxious or incurable cases. Indiscriminate spitting is strictly forbidden, and great care is taken to disinfect the *sputa*, which are received into cuspidores half full of water, collected twice daily, no handkerchiefs being permitted for this purpose. The cuspidores are disinfected by carbonate of soda, followed by steam, then by pure carbolic acid, again by steam, and finally washed with hot water, and left half full of water. The linen is disinfected by steam. Rooms vacated by invalids are disinfected by formaldehyde gas. The sanatorium is under medical control and management. The medical director is Dr. Wm. C. Bailey. There is in addition a second medical officer; also a consulting board of three and an advising board of seven, who are non-resident.

Consumptive patients are kept out of doors as much as possible. Cod-liver oil is not used, and counter-irritation only occasionally. Monthly reports are sent to physicians who send cases. For those who are fit for it there is plenty of amusement in the shape of riding, driving, mountain climbing, hunting, fishing, and out-door sports generally, besides entertainments in the lecture hall.

The institution is one year old. Last year there were 195 patients received, with four times as many guests and friends accompanying them. No charity cases are received. The length of stay is most variable. The charges are from $2 per day upwards, or $12·50 per week upwards. For patients $5 to $10 are charged for the first examination, after which a weekly charge is made of $16 upwards, including board, room, medical attendance, ordinary medicines and ordinary nursing, and meals in room if so ordered by the medical director ; but not including the treatment of un-foreseen complications or special nurse. Goats' milk is supplied free of extra charge.

THE COLORADO SANITARIUM

is a branch of the Battle Creek Sanitarium in Michigan, and was established five years ago on the western outskirts of Boulder, twenty-nine miles north-west of Denver, on a branch of the Union Pacific Railway. Boulder is an incor-porated city of 8000 inhabitants, without any factories, and is the seat of the University of Colorado. It is placed at the entrance of three beautiful cañons to the east of the Rocky Mountains, 5300 feet above the sea-level, being also protected by mountain spurs to the north. It enjoys a very fine climate throughout the year, having about 340 sunny days, a cool and pleasant summer, with the tempera-ture of the North American lakes, and a mild and genial winter. The air is exceedingly dry and bracing.

The sanatorium stands about 200 feet above the city, close to the foothills of the Rocky Mountains, and over-looks to the east and south a land of fertile prairies which grow corn and fruits in great variety. The city of Denver can just be seen from its upper floors, as well as a number of small towns and villages in the distance. To the west of the sanatorium are the snow-clad peaks of the Rockies. As the prevailing winds are from the west, and no towns or villages of any size exist on this side for 300 miles, the air is exceedingly pure. In some parts of Colorado the wind is sometimes unpleasantly strong, but at the Boulder Sanitarium strong winds are not common, and come chiefly in early spring and late autumn. The soil of the district is of sandy loam and clay. The grounds of the sanatorium cover ninety acres, consisting partly of foothills, partly of more level ground, and laid out in drives, walks and paths. There is a lawn in front of the sanatorium with flower-beds, fountains and shady trees.

The building, which is two years old, consists of a main building, with three smaller ones, all built of brick, faced with red stone, heated by steam, and supplied with lifts, electric bells and lighting. The main block, which faces east, is 110 feet long by 66 feet wide, with a large wing in the rear, and has four floors and a few attic rooms. The kitchen and dining saloon are on the fourth floor, the latter being provided with numerous large windows, with a fine view over the prairies. The three lower floors have large balconies or verandahs, the two lower running the whole length of the building on the east and south sides. The patients' rooms average 15 sq. feet and 10 feet high, and are mostly to the south or east. They are all single-bedded, with independent ventilation, and a separate air shaft to each. In addition to these there are various parlours and common rooms and two large suites of rooms for treatment, including inhalation rooms, bath and douche rooms for ladies and gentlemen, and a gymnasium where Swedish movements and various gymnastic exercises are carried out under a trained superintendent. The decoration and furniture are simple and cleansable, with no unnecessary hangings or carpets. The engines and machinery for heating and lighting are in a separate building. The water supply, which is from melted snow, comes in closed pipes for ten or fifteen miles down the mountain side. The sewerage is said to be good.

The institution is conducted by an association, controlled by a board of seven, of which the president is Dr. W. H. Riley, Professor of Diseases of the Mind and Nervous System in the University of Colorado, who was for fifteen years attached to the Battle Creek Sanitarium. With him are three other physicians, and a staff of trained nurses and attendants. The institution is not a money-making concern, and declares no dividends. Those connected with it receive only nominal salaries, or none at all. Patients with all kinds of ailments are admitted, of both sexes. The majority pay from $13 to $35 per week, including rooms, board, medical attendance, and "two treatments daily by a skilled attendant". Some are received free of charge. There is accommodation for 100 patients, but, owing to the large number of applications, it is proposed to enlarge the establishment. Last year (1897) about 700 were treated, and $8000 worth of charitable work was also done. The poorer patients lodge in the cottages, where the expenses are lower. Of consumptive patients all stages are admitted so long as there is a reasonable chance of cure ; others being advised to return home. The average length of stay is about three months. Treatment is very systematic. At first it is often found advisable to keep patients at rest. Later on various exercises are adopted to expand the chest and restore the balance of muscular development. Massage, electricity, Swedish movements, breathing exercises, inhalations of medicated vapours from a nebuliser under compressed air, hydrotherapy of various kinds, are all employed. There is systematic instruction in mountain climbing by a trained attendant for those in suitable condition. All drugs likely to be useful are employed. Cod-liver oil and creosote carbonate are occasionally used. For the *sputa* handkerchiefs are not allowed, and patients are forbidden to smoke or spit on pain of instant dismissal. They are provided with paper cuspidores in metal holders. The former are burnt and replaced every morning. Most of those who are treated for consumption receive signal benefit. Dr. Riley informs me that of early stage cases 80 to 90 per cent. are completely cured.

MANITOU PARK,

in Colorado, which has many natural advantages, is visited by a certain number of consumptives, as well as by other health- and pleasure-seekers. In the absence of regular medical officers, however, it can scarcely be regarded as a sanatorium.

ST. MARY'S SANITARIUM,

at Pueblo, California, is mentioned by Beaulavon (*Revue de la Tuberculose*, Dec., 1896). I have been unable to obtain any information about it.

Pueblo is a rising manufacturing town of 35,000 inhabitants, standing at an elevation of 4700 feet on both sides of the Arkansas river, which is a muddy, rapid-flowing stream. The soil is of clayey loam, "caking to the hardness of brick under the hot summer sun, dusty under the influence of a strong wind, muddy and tenacious after heavy rain or snow. From late September to March the climate is usually all that can be desired . . . a season of almost perpetual sunshine and moderate temperature. The spring months are more doubtful on account of occasional dust-storms and parching winds. The summers are very hot."[1]

THE BELLEVUE SANITARIUM,

at Colorado Springs, has accommodation for fifteen patients, but is shut up for want of funds.[2] The climate of the place is admirable. Dr. S. E. Solly has acted as medical officer. Colorado Springs is at an elevation of 5280 feet above the sea-level.

WHITE GABLES SANITARIUM,

or St. Mary's Sanitarium, is at Boerne, Kendall Co., S.W. Texas. The town of Boerne, which was originally a colony of Germans, is on the river Cibolo, about thirty miles northwest of San Antonio, with which it is connected by rail. It contains about 700 inhabitants, and has no factories except-

[1] S. E. Solly, *A Handbook of Medical Climatology*, London, 1897, p. 262.
[2] Beaulavon, *Revue de la Tuberculose*, Dec., 1896; private letter from Dr. Solly.

ing one cotton gin, and is surrounded by very hilly country with deep gulches. The climate is a very good one for consumptives; average rainfall, 26 inches; mean temperature in January, about 56°, in July about 89°, with usually a pleasant breeze in summer. The soil is a black loam over limestone.

The sanatorium is situated 1428 feet above the sea-level, in ten acres of ground, which is wooded to the north and east, and laid out as a flower garden and shrubbery to the south. It is sheltered from the north by a range of hills, and is open to the south.

The building is arranged round three sides of an open courtyard, the main block of three floors being at the western end, with a one-storey projection on the north side of the courtyard, and a separate block to the east containing chapel, kitchen and laundry. The main block has four rooms on the ground floor, in addition to the office, nurse's room and vestibule; on the next floor are a reading-room, small kitchen, and seven other rooms; on the top floor a nurse's room, two large wards and two other large rooms. Along the western side of the block is a large verandah, with a balcony above it; along the lateral prolongation is another long verandah. There are water-closets on each floor, and a bath-room at the eastern end of the ground floor. The heating and lighting are somewhat primitive, and some of the rooms have no chimney. There is accommodation for twenty-five patients; the bedrooms are 10 feet high, and vary from 10×12 to 16×16, and face in every direction.

The sanatorium belongs to the Sisters of the Incarnate Word (R.C.); it was originally an old stone building bought by Dr. Wm. Miller, who acted as medical officer until last year, when it passed medically under the management of Dr. A. H. Davidson, and was completely reconstructed. The number of nurses varies according to need.

Consumptives in every stage are admitted, as well as sufferers of both sexes from other ailments. Infectious

8

cases are not admitted. Patients spend much time in the
open air. The douche is not employed. Cod-liver oil, with
or without maltine, is often given. Paquin's serum was
tried in some cases, but without much success. The
sputa are disinfected by " chlorides," spitting flasks being
used indoors and handkerchiefs out of doors. Linen is
disinfected by boiling.

The charges are from $25 to $90 per month.

I am indebted to Dr. Miller for an account of his system
of managing the institution. Mainly intended for consumptives, it was necessary also to provide rooms for other
medical and surgical cases, which were placed in a different
part of the building. Acute and advanced cases were admitted to the institution, but hopeless cases were sent home.
Early and quiescent (first and second stage) cases were not
retained in the building, but sent on to one of a number of
cottage sanatoria in the surrounding country, which were
gradually organised by Dr. Miller. These consisted of a
number of detached cottages of two and three rooms, well
built and ventilated, with a central dining hall and sitting-
room, all comfortably furnished. Five such sanatoria were
built up under his supervision, with accommodation for
about 150 visitors, who were lodged and boarded with an
abundance of well-cooked suitable food for about £5 per
month. During 1896-7, 731 patients were under treatment
at the sanatorium and the auxiliary resorts, with the following results :—

Number of patients.	Stage.	Average number of days under treatment.	Im- proved.	Unim- proved.	Un- known.	Died.
1896, 420	First 73	22	24	5	44	14
	Second 102	31	35	12	55	
	Third 245	42	45	55	131	
1897, 311	First 52	20	29	4	19	12
	Second 69	35	21	8	40	
	Third 190	48	18	42	118	

CHAPTER XIX.

THE following institutions, which are scarcely to be classed with sanatoria, exist in the United States :—

Loomis Hospital and Dispensary, New York.
St. Joseph's Hospital for Consumptives, New York.
Brooklyn Home for Consumptives, New York.
ˋ The Montefiore Home for Chronic Invalids, New York.
House of Rest for Consumptives (Prot. Episc.), 1831 Anthony Avenue, Tremont, New York.
The Rush Hospital for Consumptives, Chicago.
The Cullis Home for Consumptives, Boston.
The Channing Home for Consumptives, Boston.
The Free Home for Consumptives, Dorchester, Boston.

The Hospital for Diseases of the Lungs, Philadelphia, and the Massachusetts State Hospital for Consumptives, however, which appear to base their treatment on hygienic methods, and to be suitably situated for the purpose, have been included among sanatoria, and described at pp. 94, 95 and 96.

ST. JOSEPH'S HOSPITAL,

in New York City, although not strictly a sanatorium, merits a brief description, as it is probably the largest consumption hospital in the world. It is owned and managed by the R.C. order of the Sisters of the Poor of St. Francis, who also have a number of other hospitals in various parts of Europe and America. It is open to the poor, irrespective of race, nationality or religion, and also

(115)

admits patients for a small payment into small wards and private rooms. Consumptives in all stages are admitted, but are distributed into different wards according to their stage and condition.

The hospital is situated in the suburbs of the city, between 143rd and 144th Streets and St. Ann's and Brook Avenues, occupying the whole block. It is surrounded by a garden, and looks out on the south and east on to open park grounds. It is a large pile, built ten years ago, with projecting wings on five floors including basement, with a south aspect, and contains accommodation for 365 patients. There are five large wards containing from twelve to sixteen beds on each floor, all looking south ; and at the eastern and western ends of the buildings other small wards for from two to eight, and small private rooms for one patient each. The walls are of plaster or covered with glazed paint ; the floors of varnished hardwood. The kitchen and laundry are in a separate wing. The water supply, sewerage, and gas lighting are connected with the city systems. Besides the garden, small sheltered balconies at the ends of the building are used in all weathers for the open-air treatment, but patients are drafted as soon as they are fit to one of the country sanatoria. Both sexes are admitted ; the men on the first and second, the women on the third and fourth floors. In the private wards $5 per week is charged, in the private rooms $10; but most of the patients admitted are too poor to pay anything. Last year (1897) 1500 cases were treated, and about 1000 applicants were refused for want of room. One of the cases had been under treatment for eight years.

The treatment adopted is rest in the open air ; cod-liver oil and creosote if well borne ; attention to the digestive organs, which is rightly considered to be of first importance ; tuberculin, aseptolin and similar remedies in selected cases; counter-irritants if needful, but not as routine treatment.

For expectorations handkerchiefs are not allowed. Every patient must use his own spitcup or small cuspidore ; and the *sputa* are not allowed to dry until they are destroyed by boiling water and chloride of lime. Rooms, floors, and walls are cleaned with moist mops and cloths.

There are twenty-seven nurses in the building. The medical staff consists of a physician-in-chief (Dr. C. M. Cauldwell, who kindly furnished the above-mentioned particulars), ten visiting physicians, and a house physician.

THE NEW YORK HOSPITAL AND DISPENSARY FOR CONSUMPTION

is under the same management as the Loomis Sanitarium, and was started in the spring of 1894 in a small four-storey house at 230 West 38th Street. It is intended for incurable and dying cases of consumption, and contains two wards; one for men and the other for women, with twelve beds in all. There are two visiting physicians and a house physician. The hospital is to be moved to a larger building.

THE HOUSE OF REST FOR CONSUMPTIVES

in New York has no separate sanatorium, but possesses sixty-two endowed beds in St. Luke's Hospital, in which patients in an advanced stage may be treated.

THE MONTEFIORE HOME FOR CHRONIC INVALIDS

is an institution founded and maintained in New York mainly by Jewish munificence. It is not entirely devoted to consumptives, but has two wards with thirty beds each for male consumptives and one ward of forty-two beds for female consumptives. It is primarily intended for those who are poor but are not suitable for treatment in hospital owing to the chronic nature of their ailments. Consumptives who are admitted (who need not be Jews) may stay there for the rest of their lives, but some who recover their ability to work leave of their own accord, or are (if men) sent to the country sanatorium described at p. 92. The home was built in 1887 on the Grand Boulevard about 300 yards from the Hudson river. Attached to the home is a park with large tents for use during rainy or windy weather. For weaker patients broad piazzas are available. The home itself is a fine building of red brick and granite in Italian style with a central block and two projecting wings, which together enclose an open courtyard. The centre is on five floors, including attics, the wings on four floors. In the rear between the two wings are the kitchen and laundry, the dining-room for 300 being over them, and accessible from the main hall in the centre building. In addition to the wards already mentioned, there are bed-rooms for from two to sixteen patients mostly to the south, very few to the west and north. The larger bedrooms (eight to sixteen beds) have six to ten windows on N., E. and S. There are also two very large sun rooms mainly for the winter months, a smoking-room, and large syna-gogue. The rooms are painted with a light-coloured oil paint, and have rounded angles. The heating is by steam ; the lighting by gas and electric light ; the ventilation is said to be good ; sewage goes into the Hudson river ; water supply is from the city supplies.

For disinfection carbolic acid is liberally used. Linen is disinfected in a dry hot air disinfector. Handkerchiefs are not allowed. The rooms are periodically fumigated and repainted. The douche is used ; cod-liver oil and maltine liberally given ; creosote in small doses.

The male wards have two nurses each, the female ward has three nurses. There are three resident medical officers, to the senior of whom (Dr. Joseph Fränkel) I am indebted for these particulars. Twenty-three other physicians have professionally visited the institution. During 1897, 158 were admitted and 493 treated, 189 being for phthisis. The average of admissions for the previous five years was 241.

THE BROOKLYN HOME FOR CONSUMPTIVES

was founded in 1881 under the name of the Garfield Memorial Home as a home for invalids, especially consumptives, who were not admissible to hospitals owing to the chronic nature of their ailments. It is a purely benevolent and non-sectarian institution and has two sections, one of which is attended by homœopathic physicians. A children's ward has recently been added. Both sexes are admitted free of charge, and patients in any stage, even the dying. During 1897, 236 patients were under treatment, eighty-four of them in the homœopathic section. Of those treated, eighty-five died, thirty-nine left improved, eighty-five remained in the home.

CHAPTER XX.

A SANATORIUM for consumptives of the poorer classes has been established at Alland in connection with the hospitals of Vienna. Count Batthyáni and Prof. Korányi are at the head of a society for the establishment of a similar sanatorium for the poor of Buda-Pesth. A beautifully wooded site has been chosen for this purpose not far from Buda-Pesth on the right bank of the Danube. There is also a project for erecting a sanatorium for Reichenberg in the Riesengebirge, and a similar movement at Graz under Dr. Kraus. The Minister of the Interior has had under consideration a proposal for the appointment of a special sanitary inspector for consumptives alone.[1] A number of open health resorts also exist which are frequented (mainly in summer) by paying consumptive patients ; but no closed sanatorium exclusively for this class of patients has yet been opened.

THE ALLAND SANATORIUM

was founded in 1894 and opened in 1897, mainly through the exertions of Prof. Schrötter, of Vienna, aided by Dr. Conrad Clar and Hofrath Christian Lippert. An abortive attempt was made in 1883, but it was not until ten years later that a syndicate was formed as the result of a fresh appeal, and the necessary funds subscribed, mainly by inhabitants of Vienna. The society entrusted with the erection of the sanatorium received a donation of 10,000

[1] *Heilst. Corresp.*, 1st Aug., 1898.

FIG. 30.—ALLAND SANATORIUM.

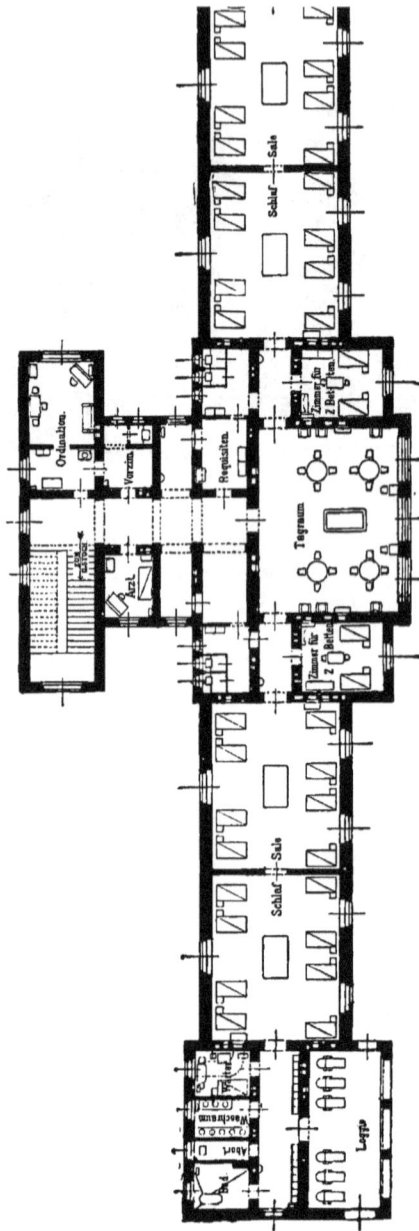

FIG. 97. AFLAND SANATORIUM. FIRST FLOOR.

FIG. 92. ALLAND SANATORIUM. SECOND FLOOR.

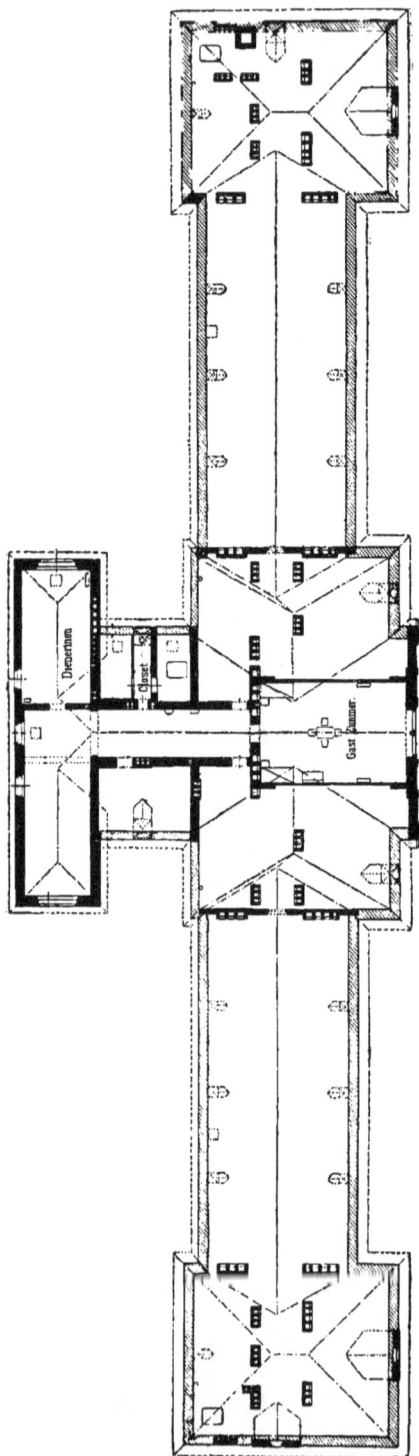

Fig. 9.—Alland Sanatorium.—Attic Floor.

florins from the Emperor and Empress, together with others ranging from 100 to 100,000 florins. The site, which is of unusual beauty, is said to have cost over 62,000 florins, the various buildings nearly 362,000 florins, and the water supply another 18,000 florins.

The sanatorium is situated in a valley in the Wienerwald, about 16 kilometres (10 miles) to the west of Baden, 27 kilometres (16¾ miles) from Vienna, and about a mile from the little town of Alland. The grounds, which have a southerly slope, cover an area of 76½ hectares (about 190 acres), and consist of woodland, meadows, and cultivated land in about equal proportions. The lowest point stands at an altitude of 400 metres (1310 feet) above the sea-level, the sanatorium itself is at 430 metres (1411 feet), and the highest part of the property reaches the top of the adjacent mountain, 680 metres (2230 feet) above the sea. Mountains exist to the east, north-east, and north-west, ranging from 485 to 680 metres (1591 to 2230 feet) above the sea-level, so that there is an absence of strong wind. Dust is also seldom noticed, and no factories or dense clusters of houses exist in the neighbourhood to sully the purity of the air. The soil is mainly of limestone, with a certain amount of clay. There was some difficulty in providing an adequate water supply, as all the streams in this region run northwards from the hills. To meet this difficulty a local land-owner gave the sanatorium the springs which exist on the other side of the hill, and an aqueduct was made with iron pipes, provided by the Archduke Albrecht, to bring the water (which is pure and abundant) round the hill to the sanatorium.

This consists of a main building, with separate kitchen block, and of other necessary buildings which are separated from the former by a little hill. The main block (fig. 9), which looks to the south, has three upper storeys, and contains on the ground floor in the centre a large day room 11 metres long (36 feet). On either side of this is a bedroom with two beds, and behind these the space for a lift, a

store room, two water-closets; and in a projecting pavilion
the bath and douche room, boot room, a large inhalation
room, and the main staircase near the principal entrance.
On either side of the centre are two large communicating
dormitories for eight beds each, and beyond these the
wings, which are occupied by day rooms in front, while the
back of each contains a nurse's room, lavatory, water-closet,.
and bathroom. This lavatory and bathroom are intended
for daily and weekly ablutions, while hydrotherapeutic
applications and the preliminary cleansing on admission
take place in the rooms near the main entrance. The
basement, owing to the rapid slope, is above ground in
front. It contains in the centre a large vestibule with
staircase, and rooms for the committee and for Jewish
worship. Laterally are the temporary chapel and the
chaplain's quarters, balanced by four rooms for the nursing
staff. In the wings on this floor are two more day rooms,.
and behind them the furnaces for heating the bath water.
A fresh-air gallery runs along the south side of the building, with a glass winter garden in the centre.

The upper floors are similarly planned to the ground
floor, with a large day room and two small bedrooms in the
centre, two large dormitories on each side, and other day
rooms, nurses' rooms, water-closets, bathrooms and lavatories in the wings. The central northern pavilion on these
floors contains the quarters for the medical officers, as well
as consulting rooms. The centre of the building has an
additional storey, of which the front forms a spare room for
medical men who may come to study the methods of treatment, while the back contains the servants' quarters. There
is altogether accommodation for 108 patients in the building, twelve in the smaller rooms, and ninety-six in the large
dormitories. The administrative block, however, has been
arranged for 300, and the grounds could easily accommodate five or six times the present number. Future additions
are to be in the form of scattered buildings in the park. In
the present building the rooms are 4·70 metres (15½ feet)·

high, 40 cubic metres (1412 cubic feet) being provided in the large dormitories, and 45 (1590 cubic feet) in the smaller bedrooms. The corners are rounded in every room.. Floors are partly of tenazzo, partly tiled, mostly covered with linoleum on a basis of plaster of Paris. The walls are washable and undecorated. Furniture in the bedrooms consists of iron bedsteads, with plain wooden tables and chairs, and open iron pedestals with glass tops. The windows are large, mostly with two hinged upright sashes. and an upper transverse sash. Ventilation is by means of windows, ventilation shafts and openings under the windows. The building is heated by low pressure steam and lighted by electricity. There is an electric heater on every floor to heat water, etc., during the night. The rain water is carried into a pond on the way to the stream ; sewage and waste water into settling tanks, where they are treated with milk of lime, and after precipitation clarified by filtration.

Coming next to the administrative block, this is placed to the east of the main building, on a lower level, and communicates with it by means of a covered corridor. It contains a cellar storey, with furnaces, etc. ; above this a small kitchen, more cellars, and the quarters of the women servants and gardener ; and on the next floor the large dining saloon, sewing room, kitchen, scullery, etc. The remaining buildings are at a distance from the main block, and consist of laundry, engine-house and disinfector, mortuary, laboratories, cowsheds and stables, together with the residence of the medical director, Dr. Alex. Ritter v. Weissmayr.

Patients are chosen at the general hospitals of Vienna, and are exclusively consumptives of the male sex in remediable stages. After a probationary period of three weeks, they usually stay three months. They are allowed in certain cases to do gardening and other suitable work. They pay 1 florin per day, or 2·50 in the smaller rooms. The parish or sick fund pays for the poor. There are five

nurses in the establishment. The fresh air treatment is
carried out in the fresh air galleries, or in the day rooms
in the wings. The douche is only applied to robust
patients. Cod-liver oil is not given. Drug treatment consists mainly of codeia, phenacetine, etc. *Sputa* are received
into Dettweiler's flasks, or at night into spitcups of *papier
mâché*. The contents of these vessels are mixed with peat
mould and burnt. Linen is disinfected by steam; rooms,
in case of need, by formalin vapour.

The sanatorium may be reached by the Southern Railway or the South-Western Railway. In the first case
there is a drive of one and a half hours to Schwefelbad,
three-quarters of an hour by rail from Vienna; in the
second, a drive of three-quarters of an hour to Altenmarkt
or Weissenbach, one and a half hours' journey from Vienna.

The Sanatorium of Marilla Völgy,

in the South Carpathians,[1] stands at an elevation of 714 metres
(2342 feet) above the sea-level. It is well sheltered on all
sides against wind, and is isolated by beautiful woods. Dr.
Hoffenreich, the medical officer, also receives non-phthisical
patients. The place is lit with electricity, and has covered
verandahs for the fresh air treatment. The diet is very
strict.

The Sanatorium of Uj-Tatra-Füred,

or Neuschmecks, also in the Carpathians, is under the care
of the well-known Dr. v. Szontagh, and is placed 1004
metres (3294 feet) above the sea-level, in the highest part
of the village of Gerlach. The soil is of clay; the winter
dry and cold, the summer cool, of medium humidity. Fogs
are rare; and the absence of any glacier in the neighbourhood is also an advantage. Mountains of 3000 metres
(9840 feet) shelter the sanatorium to the north, and there
is very little dust or wind from any quarter. The sana-

---

[1] I am indebted to Kuthy (*A Tüdővész Szanatoriumok*, Buda-Pesth, 1897)
for the following descriptions of Hungarian sanatoria.

torium has a southern aspect, and has a ground floor and two upper floors. The eastern end consists of four projecting square towers with pointed tops, connected by a lower central portion. Great pains have been taken to render the building thoroughly dry. There is central heating by warm air, and ventilation of the same kind in cold weather. In the rooms for one patient the air can be renewed fifty times per diem. The sanatorium communicates with the thermal bathing establishment and the dining saloon by means of a covered corridor. It is a great drawback that, owing to this arrangement, and to the proximity of several other bathing establishments, patients cannot be completely supervised. There are several balconies on the south and west sides of the building, and a large winter garden. The rooms are well lighted; there are double windows, hot-water pipes in the corridors and staircases, and a separate stove in each room.

It is mainly a summer resort, and not exclusively reserved for consumptives, who are received in the early stages alone. Dr. v. Szontagh uses massage and cold ablutions, friction with moist compresses, seldom any douches, but pulmonary gymnastics and inhalations. The *diet* is abundant, but not medically supervised. Milk and kefir are largely given.

The sanatorium is nine kilometres (five and a half miles) from the railway station of Poprad Felka, nine hours by rail from Buda-Pesth, and a little more from Breslau.

A few poor patients are received, but the institution is essentially for paying patients.

Kuthy mentions three other places in Hungary which would be suitable for consumptives, but are little frequented by them. Of these, *Feketehegy* is 660 metres above the sealevel; it is right in the country; but the diet is meagre. *Stoosz* is 670 metres above the sea, and is directed by Dr. Dezsoe Czirfusz, a specialist in consumption.

At *Keresztenysziget* (Isle of the Christians), in the province of Szeben, there is a fine establishment founded by a

society, in June, 1894, with the aid of various philan-
thropists, as a Convalescent Home, and called the "Curhaus
auf der Hohen Rinne". It is 1420 metres (4659 feet)
above the sea, in a pine wood, about five or six hours' drive
from Nagy-Szeben. The principal block contains the dining saloon and administration, besides which there are two
wings, bath-houses, and other buildings. The rooms are
comfortable, for one, two, three and four persons, mostly
with verandahs, and all capable of being warmed. There
is accommodation for sixty to seventy patients. The food
is excellent, three *meals* per day being provided. Patients
can be treated by hydrotherapy, massage, electricity, and
inhalations. The charges are 12 florins per week.

Dr. Schreiber's Alpenheim,

at Aussee, in Styria (700 metres or 2300 feet above the sea-
level), is also frequented by consumptives. I regret that I
have been unable to obtain any information about this
sanatorium.

Consumptives also visit *Meran* in the Austrian Tyrol for
treatment by hygienic methods. This place is situated on
the southern slopes of the Alps, at an altitude of 1050 feet
above the sea-level. The climate is dry and bracing, with
an average of seven snowy and fifty-two rainy days. The
mean winter temperature is higher than at Montreux, but
decidedly lower than at Torquay. There is a fine Kurhaus,
with a number of reception-rooms, and arrangements for
medicated baths, pneumatic chamber, grape cure, whey cure,
etc.[1]

[1] Solly, *loc. cit.*

CHAPTER XXI.

THE prevention and treatment of tuberculosis has received much attention in France of late years. Several important societies have been started which have had a most useful influence on public opinion, and have led to better public prophylaxis, and directly or indirectly to the establishment of a number of sanatoria. The most important of these societies are the *Œuvre de la Tuberculose* (scientific) ; the *Ligue contre la Tuberculose* (popular); the *Œuvre des Enfants Tuberculeux* ; the *Œuvre des jeunes filles poitrinaires* ; and the *Société des Sanatoria de France*. The *Œuvre des Hôpitaux Marins*, the *Œuvre des Hôpitaux de Montagne*, and the *Assistance Publique* of Paris and other towns and cities are also concerned with tuberculosis, but do not limit their scope to this disease. The *Œuvre de la Tuberculose* holds periodical congresses and publishes a monthly periodical, *Revue de la Tuberculose*. One of the physicians attached to the *Œuvre des Enfants Tuberculeux* has also started a bi-monthly periodical called *La Tuberculose Infantile*. This society has done a most important and practical work in establishing sanatoria for children (see p. 139).

There are at present three sanatoria for paying consumptive patients in France, all established by private initiative —the Canigou Sanatorium, the Durtol Sanatorium, and the Trespoey Sanatorium. Two other sanatoria of this kind are projected by the *Société des Sanatoria de France* at Mont Bonmorin in Puy-de-Dôme and Mont Pacanaglia near Nice. The society has also bought land in the neighbour-

hood of Ajaccio (Corsica) for three sanatoria—one of which
will be on the coast; another at Vivario, at an elevation of
950 metres, for 100 beds ; the third at Monte d'Oro, at 1700
metres above the sea, also for 100 beds, only fifty of which,
however, will be kept open in summer. These sanatoria
will be partly for paying, partly for poor patients, the one
set paying for the other. The society also proposes to
erect pavilions for the reception of consumptives in various
towns, the current expenses, to be defrayed by the *Assistance Publique*. A sanatorium for the poor is being erected
at Angicourt (Oise) jointly by the Paris Municipal Council
and the *Assistance Publique*. This has been a long while
under construction, but is not expected to be open for a
year or two. It has already cost a large sum of money,
and there is a fear in some quarters that it is being built
in an unnecessarily luxurious style. Eventually intended
to accommodate 200 patients, it will at first be opened with
beds for fifty or 100. It is said to be modelled on the
Falkenstein Sanatorium, with a deep verandah along the
southern side, but with only two floors. The rooms are to
accommodate from one to eight patients each. The heating
will be by means of steam and hot water, the lighting by
electricity. There are twenty-eight hectares (sixty-nine
acres) of ground set apart for the institution. A sanatorium
is also being erected at Hauteville for the poor of Lyon. A
society of eighty medical men exists at Arcachon for the
erection of two sanatoria in the neighbourhood of that
place for phthisical adults. There are also schemes for the
erection of a sanatorium at Magny (Dép. du Rhône) and
for another in connection with the Canigou institution for
paying patients. A sanatorium at St. Symphorien started
by Dr. Chaumier of Tours unfortunately failed for want of
funds.[1] There is a sanatorium at Agnetz (Oise), which was
recently erected by the *Assistance Publique*, which receives
patients at a charge of 15s. per diem.[2] The Berck-sur-

---

[1] Möller, *Les Sanatoria*, Brussels, 1894.
[2] *British Medical Journal*, 28th May, 1898.

Mer Hospital and some of the Paris hospitals also receive paying patients at £2 10s. per month. A home or refuge for incurable cases of the female sex exists at Villepinte (Seine et Oise), and a similar institution for 400 patients has been projected near Brévannes.[1] Three of the French hospitals (Boucicaut, Laennec and Lariboisière) have set apart certain wards for the reception of phthisical patients ; 6,000,000 francs have been voted for the construction of new wards for the same purpose at the hospitals of St. Antoine, Cochin, Broussais, Bichat, La Pitié, Tenon, and a new hospital on the right bank of the Seine ; and special wards have also been opened under Prof. Grasset at the hospital of Montpellier.[2] There is a "sanatorium" at Algiers, but I believe it is of a different kind.

FRENCH SANATORIA FOR PAYING PATIENTS.

Canigou	Pyrénées Orientales	2200 feet	100 beds
Durtol	Puy-de-Dôme. .	1706 ,,	32 ,,
Trespoey . . .	Basses Pyrénées .	696 ,,	14 ,,
Mont Bonmorin .	Puy-de-Dôme . .	2630 ,,	120 ,,
(being built)			
Mont Pacanaglia .	Alpes Maritimes .	1640 ,,	120 ,,
(projected)			
St. Martin Lantosque .	Do. .	3280 ,,	?

SANATORIA FOR CHILDREN OF THE POOR.

Ormesson	Seine et Marne .	374 feet	130 beds
Villiers-sur-Marne . .	Do. .	397 ,,	220 ,,

[1] Kuthy, loc. cit.

[2] Léon Petit, Tuberculosis Congress, Paris, 1898 ; Netter et Beauvalon, Gaz. Hebd. de Méd. et de Chir., 18th August, 1898.

9

CHAPTER XXII.

The Canigou Sanatorium.

This sanatorium, which is near Vernet les Bains, Pyrénées Orientales, was the first private establishment of the kind in France, being founded in 1890 under the direction of Dr. Ch. Sabourin. Situated at an altitude of 640 to 700 metres (2100 to 2297 feet) above the sea-level, at the intersection of the valleys of the Cadi and the Tech, it has a southerly aspect, looking towards the Puig de Falgouras and the town of Le Vernet, which has been known since 1181 for its sulphur springs. Mountains protect it from wind on nearly all sides—the Canigou to the east, the Cerdagne Mountains to the north and the Perra to the south, leaving an open valley towards the south-west. The soil consists of sand and pebbles resting on gneiss and granite.

The sanatorium is situated in a park of 20 to 25 hectares (49 to 61 acres), which is well wooded with chestnut trees, acacias, oaks and evergreen oaks, and various kinds of pine trees. Palms, aloes, olive trees and cacti grow freely without artificial shelter. It was originally an "open sanatorium," in which the patients merely spent the day during the winter season, returning to sleep in hotels and apartments in the town. At that time it consisted of a series of verandahs open on one side, glazed on the other; of a dining-room, drawing-room, consulting-room, administrative and douche rooms. In September, 1896, however,

(130)

when Dr. Giresse succeeded Dr. Sabourin, it was opened for the whole year, and completed by the addition of sleeping quarters.

The present building, which is of three storeys, has accommodation for about 100 patients; a new building is, however, in contemplation. The bedrooms, which face south, south-east, west and north, have a mean capacity of 60 cubic metres (2119 cubic feet). The windows, which are wide open during the day, and partially open at night, are provided with screens near the beds, but no curtains. During the coldest part of the year (15th December to 15th January) they are heated with warm air: for the rest of the year no artificial heat is required. The passages are lighted with gas, the bedrooms only with candles. The drinking water is taken from a torrent which comes down from Mount Canigou, and is filtered through sand, pebbles and iron scoria. The sewage is carried by an abundant water supply into the river a few hundred metres below the village. There are nurses as well as servants in the establishment.

Every patient has two spitcups of porcelain, one for his room, the other in the open-air gallery, besides a pocket flask; and spitting elsewhere is strictly forbidden. The spitcups are emptied every morning, the *sputa* mixed with sawdust and burnt in the gasometer furnace. Linen is disinfected with steam by means of one of Geneste and Herscher's disinfectors. Rooms are disinfected with formol vapour. Some of the fresh-air galleries are near the main building, others scattered through the park.

The average length of stay is five to six months; patients are allowed to go home after a while, returning for other courses of treatment until they are cured. There would be great advantages if they were to stay until their recovery is assured; but they would seldom stand so long a separation from their families.

Drugs are not much used; those most employed being arsenic, phosphates and opium. The douche and the wet-sheet are only occasionally applied. Counter-irritation is employed, either as actual cautery, or in the form of sinapisms or blisters.

The charges are from 14 to 16 francs, including lodging, board and medical attendance ; drugs, etc., and laundry being extra.

The statistics furnished by the present medical director (Dr. Giresse) show 22 to 23 per cent. cured, 40 to 50 per cent. improved, 20 per cent. stationary, and 10 per cent. worse—chiefly patients who come for treatment in a late or acute stage of the disease. Dr. Giresse believes that of early cases and with manageable patients there should be from 50 to 60 per cent. cured.

The most convenient route to the sanatorium from Paris is by the Barcelona night express to Perpignan, thence to Villefranche de Conflent, which is about half an hour's drive from the sanatorium.

THE DURTOL SANATORIUM,

or the sanatorium of the Château of Durtol, is situated on very pervious black volcanic sand, 520 metres (1706 feet) above the sea-level, at a little village three kilometres from the large town of Clermont-Ferrand in Puy-de-Dôme. It has a station of its own on the line running from Clermont-Ferrand to Limoges, and is half an hour's drive from the Clermont station, or nine hours' journey from Paris. Being somewhat raised above the plains, it has a good view of the valley of Clermont and of the distant mountains of Forez. It is open to the south, protected to the north, north-east and north-west by mountains and hills thickly wooded with pine trees, and provided with numerous good roads and paths. The strong winds, which come from the north-east, are completely warded off by the woods extending the whole length of the park, which is of five hectares (twelve acres). The climate is sedative compared with the rest of Auvergne. The rainfall is regular but not abundant, and prolonged droughts are unknown. Excepting in very severe winters, the cold weather is of short duration and not very intense.

FIG. 10.—SANATORIUM DU CHÂTEAU DE DURTOL, PUY-DE-DÔME, FRANCE.   [Face page 183.

Fogs seldom reach the place. The sanatorium consists mainly of an old Louis XIV. château (fig. 10), which has been altered and added to by the proprietor, Dr. Chas. Sabourin, formerly director of the Canigou Sanatorium. There are rooms for thirty-two patients on two upper floors. Nearly all the rooms have chimneys, and the windows are kept open permanently. It is very seldom that a fire is needed, rugs and plaids, etc., being sufficient to keep the patients warm. The lighting is by petroleum. The floors are of pinewood, polished with linseed oil. The angles of the rooms are rounded; the walls papered. Dettweiler's flasks are used, the *sputa* being destroyed by boiling. Linen is disinfected by steam heat. On the departure of a patient the walls are repapered, the paint renewed, and the furniture disinfected. There are two large verandahs for the fresh air treatment. The sanatorium has its own dairy farm and its own private (R.C.) chapel. Hydropathy is not employed excepting in the form of cold bathing or friction with cold water, and so-called specifics are not administered. Dr. Sabourin attaches great importance to the digestive functions, and has obtained very striking results from super-alimentation with raw meat, meat powder, beef peptones, eggs, milk, and the like. Cod-liver oil he gives in full doses where it does not upset the stomach. Dyspepsia in his opinion often disappears promptly under the fresh air treatment. He has published a case of incoercible vomiting which ceased within twenty-four hours of starting the treatment.[1] Dyspepsia may also sometimes be prevented by the simple expedient (which Sabourin always prescribes) of gargling the mouth and thoroughly cleansing it with an alkaline water such as Vichy or Vals, as this prevents the fermentation of particles of food and the entrance of various bacteria into other parts of the alimentary tract. Many patients are quite easy to feed, but some will tax the ingenuity of their physician to

---

[1] *Gaz. Hebdomadaire*, 31st Oct., 1891.

the utmost.   Many dyspepsias disappear if patients drink
little at meal times and give up wine.   Where drugs are
needed to control cough, Sabourin depends mainly upon
opiates, which in his opinion do much good and little harm.
The fever which is due to excessive waste usually subsides
with rest in the fresh air.   That due to tubercular processes
cannot be prevented by any known drug, so that Sabourin
only gives antipyretics to prevent irregular febrile attacks
at unusual hours and those accompanied by much discom-
fort.   The latter as a rule is not present where the patient
lives in fresh air day and night.[1]

The sanatorium, which was started in 1898, is open the
whole year.   Durtol is a telegraph station, and is very near
some of the famous bathing resorts of central France.   The
charges are 14 frs. per diem, including wine, three meals per
diem and medical attendance.   The room is from 3 to 5 frs.
extra.

THE TRESPOEY SANATORIUM,

about half an hour's drive from Pau (Basses Pyrénées), is
situated in the midst of a cultivated plain, on the edge of
the valley of the Ourse, which is about 40 metres below.
Originally a private house, it was altered and adapted to
its present purpose in 1896 by the proprietor and medical
director, Dr. Crouzet.   The climate is chiefly remarkable
for the absence of wind, high wind only blowing three or
four times during the course of the winter.   There are no
forests in the neighbourhood; the high road to Pau has a
number of villas on each side with patches of woodland,
which form a protection against the west wind ; there is
similar protection towards the east.   The soil is a very
permeable sand to a depth of about 4½ to 5 metres, where a
thin layer of clay is met with.   The elevation is 212 metres
(696 feet) above the sea-level, the mean B.P. being 740 mm.
The sanatorium is on a steep hill, and has a fine view of
the Pyrenees in the distance and the river valley in the

---

[1] *Traitement Rationelle de la Phtisie*, Paris, 1896.

FIG. 11.—THE TRESPOEY SANATORIUM, NEAR PAU, S. FRANCE.

[*Face page* 135.

foreground. It stands in a park of about 7 hectares (17 acres), containing pine trees and other conifers, and laid out in paths at various inclinations, well provided with seats. Round it is meadow land.

The sanatorium consists of two buildings, the chief of which is of two storeys above a basement and ground floor. In the basement are the kitchen department and bath-room. On the ground floor the dining-room, library, drawing-room, consulting-room, and two rooms for patients. The next floor has six bedrooms, and the top floor two bedrooms for patients. There is no lift. The patients' rooms are all to the south, with at least one south window ; they vary in size from 60 to 110 cubic metres (2119 to 3885 cubic feet). Each has its own chimney and open wood fire ; ordinary French windows ; papered walls, without rounded corners ; pinewood floors, without carpet or curtains ; and varnished pitchpine furniture, without carpets, pictures, or curtains in any of the rooms. The lighting is by petroleum. I am informed that there will be electric light at the end of 1899. There is a good water supply ; sewage goes into a cesspool. Near the chief building is a fresh air gallery, with a southerly aspect (fig. 11). The annexe, which is a building of one storey, contains bedrooms for four patients. The total accommodation in the sanatorium is for fourteen patients. The sanatorium is open from 15th October till 15th May, and receives none but paying phthisical patients at an early stage. *Sputa* are disinfected by boiling with solution of carbonate of soda, and then poured down the water-closet. Expectoration is exclusively into Dettweiler's flasks or spittoons, which contain one per cent. sulphate of copper. The Thoinot system is to be introduced. Linen is disinfected once a week in the municipal steam disinfector. Rooms are disinfected after the departure of each patient by means of pulverisation with sublimate solution. This is done by employees of the city of Pau. The wall paper is of a cheap kind, and is frequently renewed.

The treatment is by fresh air, good food, daily friction with a wet horsehair glove, and as few drugs as possible.

Patients who are not feverish take frequent short walks. The diet is highly nitrogenous, and the medical director takes his meals with the patients. There are three meals a day. Cod-liver oil, specific remedies, counter-irritation, and douches are not employed.

Dr. M. Crouzet is the sole medical officer.

The charges are from 16 to 20 frs. per diem, and a fee of 20 frs. on departure for disinfection. Extras are laundry, special drugs, and nurses for the night if required.

THE SANATORIUM OF MONT BONMORIN,

although not yet opened, has been fully described by E. Marty-Martineau in *L'Indépendance Médicale* for 25th March, 1896.

It is situated near Ardes (Puy-de-Dôme), at an altitude of 2630 feet above the sea-level, on the side of a mountain, and has a fine view to the south over the rich plains of the Limagne. The sanatorium consists of three principal buildings united by slightly curved covered corridors, forming altogether a somewhat concave southern front, as at Ruppertshain (p. 253). It has behind it a park of about 500 metres depth, traversed by geometrically arranged paths and drives which unite in a *rond point*. The villa of the chief medical officer is at some distance from the main building, on the western side of the park, and is balanced by the bathing establishment on the east. The sides of the park are occupied by private and kitchen gardens. In front of the main buildings is a flower garden with a central fountain and geometrical walks and drives.

Of the three principal buildings, the middle one consists of a central block with lateral wings. The former is surmounted by a cupola, and contains on the ground floor a large vestibule with carriage drive through to the park at the back, and a drawing-room on each side. Behind are the two approaches to the grand staircase. Above the vestibule is a large concert hall, 13·50 metres square, with balconies front and back, and above this, under the dome-shaped roof, another large assembly room. The wings of the central block consist of a single row of rooms in front and corridors to the back, excepting at the ends, which are somewhat higher and have rooms back and front. On the ground floor are a series of rooms for common use—in the left wing the dining saloon (48 × 5·50) and staff dining-room (20 × 6), in the right wing a reading-room (16 × 5·50), writing-room, billiard-room (16 × 5·50), lavatory, and two large recreation-rooms (19 × 6). Behind the staff dining-room is the administrative block with kitchen department under it. The first and second floors on each side contain in front eighty patients' rooms, and behind sixteen rooms for the staff, with more under the roof. In addition to the grand staircase, there are two lateral ones, and two for the servants. There are four groups of water-closets on each floor at the back, occupying the projecting portions of the building. The lateral buildings are somewhat similar, but without a cupola. They contain on the ground floor a vestibule leading to staircases, waiting and consulting rooms, laboratory and doctors' rooms ; and on the first and second floors forty [1] bedrooms for the patients and sixteen for the staff. Here also there are four staircases, of which two are for servants, and two groups of water-closets on each floor. An arched fresh air gallery runs the whole length of all three buildings level with the ground floor, the intermediate portions being glazed and 50 metres long. The bedrooms are all 4 × 4½ metres and 3·40 high (13 feet × 14⅔ × 11), containing therefore 61·30 cubic metres (2165 cubic feet) of air space. Each has a chimney and a balcony, a large window in three parts, the upper of which has a ventilator which cannot be completely closed : above the balcony is a ventilating inlet which is larger inside

---

[1] According to the description it would appear that there are forty patients' bedrooms in each lateral building, but I believe there will only be twenty on each side.

than out, and above the upper window pane is another ventilator. The walls are painted, decorated and varnished. The floors are covered with asphalt or mosaic, and slightly sloping to aid in drying them. There are nowhere in the building any projections, angles or corners, or any surfaces likely to accumulate dust. There are no curtains or draperies, and no pictures, only mural paintings. The furniture is also designed on similar principles: a metal bedstead with movable mattress, a pedestal, two leather sofas, and a reclining chair of bamboo. There will be room for 120 patients.

The Sanatorium at Mont Pacanaglia,

four and a half miles from Nice, is intended to be built on the same plan as the one at Mont Bonmorin. The site which has been secured affords a magnificent view over the bay of Villefranche, and is 1640 feet above the sea-level. Both sanatoria will be well sheltered from north winds. The winter climate of Mont Pacanaglia and the summer climate of Mont Bonmorin are said to be exceedingly equable, with a range of temperature from 18° to 22° C.

Both of these sanatoria are being erected by the *Société des Sanatoria de France*, under the direction of Dr. S. Bernheim, who described them at the Int. Congr. for Climatology held at Clermont-Ferrand in 1896.

St. Martin Lantosque,

near Nice, also has an establishment for the reception of consumptive patients, which stands 1000 metres (3281 feet) above the sea-level. It is, however, scarcely more than a summer resort, and is not well supported (L. Petit, *loc. cit.*, p. 253).

CHAPTER XXIII.

A VERY complete organisation exists in France for the care
of consumptive children. This is directed by a society
called *L'Œuvre des Enfants Tuberculeux*, which was.
founded by private initiative at the end of 1888, and
recognised as of public utility by a decree of the French
Government on 18th January, 1894. This society, unique
in French annals, is entirely supported by private subscriptions and is intended to promote the foundation of
hospitals, asylums, sanatoria, dispensaries, agricultural
colonies and other establishments for the gratuitous treatment of tuberculous children of poor patients, irrespective
of creed or nationality. The society (which has a council
of twelve and is under the scientific direction of a medical
committee of twenty members) now has a dispensary in
Paris, two hospitals or sanatoria in the country, and several
small agricultural colonies for the training of those who
have been cured in the hospitals, thus providing an unbroken series of steps whereby a tuberculous child may be
cured and afterwards fitted for a healthy open-air occupation. The dispensary is in the Rue Miromesnil, at its.
junction with the Rue la Boëtie. It was rebuilt in 1890,
and is extremely well suited to its purpose, having light
and lofty rooms, with walls and floors which can be readily
cleansed. During 1897, 1652 patients were seen at the
dispensary, out of whom 1474 were tuberculous or pre-
disposed to tuberculosis. All adults are sent to other
clinics, as well as children suffering from non-tuberculous

(139)

ailments ; and an attempt is made by systematic cross-
questioning to prevent the abuse of the charity by those
who can afford to pay a fee. Dr. Derecq and Dr. George
Petit are the physicians.

The hospital of Ormesson receives consumptive boys up
to the age of twelve, those from twelve to sixteen being
sent to the hospital of Villiers. None but consumptive
children are admitted, and no selection is made, those in
every stage being accepted.

THE ORMESSON HOSPITAL

was started in an ordinary house at the end of 1888 ; enlarged to one hundred beds in 1890, with the help of the
wood from the Exposition Universelle : and completely
rebuilt in 1896 with 130 beds, so that it now covers an
area of 1956 square metres (21,000 sq. feet). It is situated
on a breezy plateau overlooking the valley of the Marne,
in the little village of Ormesson, just above Champigny.
The ground slopes towards the south, and the building,
together with the high walls surrounding the grounds,
afford a little protection against high winds ; but the situation is breezy rather than sheltered. There are 5¾ acres
belonging to the institution, which are partly devoted to
playgrounds, partly consist of garden, cultivated and grass
land. The soil is calcareous, and rapidly dries after rain.
The building (fig. 12) is 114 metres (374 feet) above the sea-
level. It is of one storey, and consists of a large central glass-
covered winter garden with huge windows to the south,
and forming with the administrative offices a central axis,
on either side of which are two parallel pavilions, making
with it a letter H. One of these pavilions forms the refec-
tory, and has at the western end the kitchen block. The
other three form dormitories, each for about forty beds,
and having at the free ends lavatories, bath-rooms and water-
closets. The latter are arranged in radiating compartments, so that a number can be supervised by one person.
They are isolated by the lavatories and bath-rooms, which

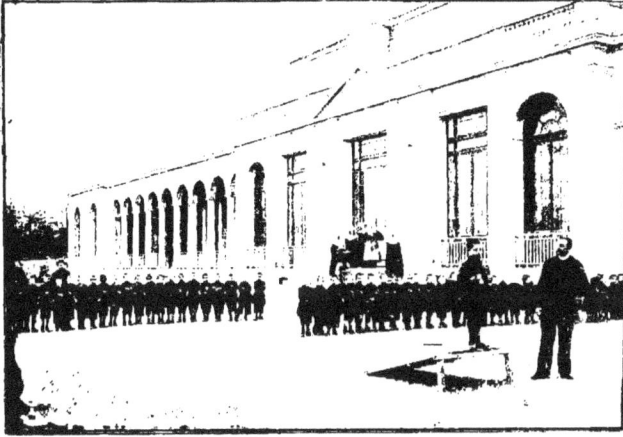

FIG. 12.—THE ORMESSON HOSPITAL. [*Face page* 140.

can be ventilated from end to end. The sisters' rooms lead off the dormitories. The beds are of enamelled iron; floors are of wood in the dormitories, of tiles in the dining hall; the walls everywhere washable, with rounded angles. Windows are large, of the usual French type, in three parts, with the tops rounded. The whole place was exquisitely clean when I visited it. Along the south side of the building is an open verandah on each side of the projecting winter garden. This is somewhat narrow; and there are, I believe, no other open-air shelters for bad weather; but the winter garden is very large and well-ventilated, and is probably sufficient for the purpose. The children in the playground are placed under the supervision of two men; indoors they are under the care of the sisters. Drugs' are but little used, reliance being chiefly placed on fresh air, good food and cleanliness. The results of treatment have been very satisfactory. In 1897, 34 per cent. were apparently cured, 30 per cent. more improved. Out of fifty-three who were apparently cured, fifteen had very serious and extensive lesions on admission. Dr. Jaoul is the medical officer.

THE VILLIERS-SUR-MARNE HOSPITAL,

which has accommodation for 220 boys, is about three miles from that of Ormesson, and stands on the same high plateau, 121 metres (397 feet) above the sea. It is in the midst of cultivated fields and gardens, at the extreme end of the village, and has 9½ acres of ground belonging to it. It consists of three blocks arranged along one line nearly 400 feet long, forming a long two-storey building. The southern side is separated from the road (which is not a main road) by a large gravelled courtyard. On the other side of the road is a patch of garden bounded by trees. The grounds on the north are occupied by playgrounds and meadows; they rise towards the fortifications of Villiers, with wooded crests in the distance. As in the case of the Ormesson Hospital, there is very little shelter. The soil is the same. The building was founded in 1891, but was completed in 1896 by the addition of the Pavillon des Enfants de France. The central block contains on the ground floor two large day rooms, the library, drug room, and doctor's room ; and has in front along the south side a verandah with stone arches. On the floor above is a dormitory for the more serious cases, with a single row of beds. It is incompletely divided into separate rooms, and opens in front into a covered balcony with stone arches. Attached to the dormitory are two rooms for the nurses.

The left or eastern wing, called the Pavillon des Enfants de France (fig. 13), is a large hall containing 10,000 cubic metres of air, or 120 cubic metres (4238 cubic feet) per head, with an ogival roof rising to 12 metres (39 feet) above the ground, divided into a central and two lateral naves, and partly composed of glass. This is supported by graceful iron columns, and provided with a gallery 6 metres (19½ feet) wide, which runs all the way round. The building is of brick and stone, coated internally with white impermeable enamel, the angles being also rounded. The centre of the ground floor is tiled, and contains the ventilating openings and ozonisers, and is furnished with a few chairs and tables. The part underneath the gallery is slightly raised above the ground level. This and the gallery have varnished boarded floors, and are provided with a single row of blue enamelled iron beds radially arranged, together with other simple furniture. The windows are very large, being 2 metres wide and 4 metres high. They consist of two parts, of which the lower is a sashed window (English fashion) ; the upper a French window in four pieces, each of which can be separately opened, two moreover being perforated. The hall rests on a basement 3 metres deep impermeable to moisture, and containing the heating and ventilating apparatus, the bacteriological laboratory and electric lighting machinery.

In this basement, beside the calorifer is a reservoir for a medicated solution of creosote, turpentine and eucalyptol,

FIG. 13.—THE VILLIERS-SUR-MARNE HOSPITAL.—PAVILLON
DES ENFANTS DE FRANCE.

FIG. 13*.—THE VILLIERS-SUR-MARNE HOSPITAL.—SOUTH FRONT.

[Face page 142.

which serves to medicate and purify the incoming hot air. The fresh air from outside passes through metallic gauze either directly into the hall or past the warming apparatus. The hot and cold air inlets are placed side by side in the embrasures of the windows. Round the galleries at the head of the beds are a series of tubes with trumpet-shaped ends, which supply ozone produced by the dynamo. The windows also allow of the direct entrance of fresh air. The vitiated air is carried off through holes in the roof with the help of a jet of steam in a turret. In this way at least 200,000 cubic metres (over 7,000,000 cubic feet) of air are provided in the hall per diem. Eighty children can sleep in this hall, and two sisters, one for each floor, can readily look after them. At the northern end of the Pavillon des Enfants de France on the ground floor are the dressing-rooms, lavatories, and bath-room. The latter forms half a circle, and is incompletely divided by radiating partitions 2 metres high, separating ten white enamelled iron baths. Hot and cold water are laid on in the bath-rooms and lavatory, which are arranged in English fashion, and (like the rest of the building) are heated with hot-water pipes. The walls are covered with Dutch tiles, and have rounded angles; the floors are of mosaic; the wash-basins of white stoneware on white marble. Uniting the Pavillon des Enfants de France to the central block is the grand staircase, behind which are water-closets and lavatories on each floor.

The western wing is older than the eastern, and contains on the ground floor the kitchen department and chapel. On the upper floor are the quarters for the *personnel*.

The hospital is lighted with electric light. Filtered and sterilised water is laid on in the sick ward.

To the west side of the hospital grounds is a long line of one-storey buildings, beginning with the lodge, and including the laundry, steam disinfector, linen store, clothing store, provision department, and workshops for repairs. They are large enough to serve all three establishments

—the dispensary and the two hospitals at Ormesson and Villiers. Dr. Vaquier is the medical officer. The treatment at Villiers is of the same character as at Ormesson. During 1897 there were ninety children who were treated at Villiers. Out of these, twenty-three left the place apparently cured ; thirty-one improved (two of these left the place ; some others are nearly cured) ; twenty-four remained stationary ; four are in a grave condition, and eight have died. This gives a percentage of 25·5 cured ; 34·4 cured or nearly cured ; 8·8 dead. Most of the latter entered at a late stage of the disease.

Altogether 1068 children have been treated at the two hospitals from 1889 to 1897 inclusive ; 317 of these (or 29·7 per cent.) have been cured, 47 (or 4·4 per cent.) have died. The average duration of treatment has been 230 days.

The results at Ormesson are better than those at Villiers, probably because the patients more often enter at an earlier stage of the disease, and because under suitable conditions the younger patients are more readily curable. There is year by year a greater proportion of apparent cures, although all phthisical children without selection are sent to the hospitals.

The agricultural colonies connected with the movement are at present two in number—one of 12 hectares (29 acres) at Noisy-le-Grand, not far from Villiers ; the other at Tremilly in the Haute Marne. At each of these colonies there is a medical man in charge. In addition to these, a few patients have been sent to a farm at Rougemont (Doubs). Altogether eighteen patients have left the hospitals apparently cured. Every one has remained in good health ; six have been drawn for conscription, and five accepted, the sixth being rejected for a reason quite unconnected with tuberculosis.

OTHER SANATORIA.

Another sanatorium was also for a time in existence at Touraine, near Tours (St. Radégonde), for delicate and

consumptive boys of the poorer classes, founded and supervised by Dr. Chaumier.   It was at this establishment that
creosotal was first clinically tested in phthisis.   The sanatorium failed for want of funds.   A similar establishment
(Villa Lapierre at St. Symphorien) also failed.   Over twenty
seaside and mountain sanatoria also exist in France, with
between 2000 and 3000 beds, mostly for children with
scrofulous or consumptive tendencies, but not confined to
these ailments.   They resemble our own convalescent homes
rather than true sanatoria for consumptives.

10

CHAPTER XXIV.

DURING the last twelve months special arrangements have
been made at three of the Paris hospitals (Boucicaut,
Lariboisière and Laënnec) and at the Montpellier hospital
for the hygienic treatment of consumptives.

At the Boucicaut Hospital, which was founded by the
late proprietors of the Bon Marché, two out of five wards
have been set apart for this purpose since November, 1897.
There are twenty-one beds for male consumptives, eighteen
in one large ward on the ground floor, one single-bedded
room reserved for the staff of the Bon Marché, and
two other beds; and fourteen beds for women, of which
ten are in one large ward on the ground floor, and two
single-bedded and one double-bedded room on the first
floor, all three smaller rooms being also reserved for the
Bon Marché. This section of the hospital is under the
care of the senior physician, Dr. Letulle. According to
a report given at the Tuberculosis Congress at Paris, in
August, 1898, the beds reserved for the Bon Marché staff
have remained almost unoccupied; but the other thirty-one
beds have been constantly in use. During eight months
125 patients have been under treatment, mostly suffering
from the more advanced stages of the disease; with the
result that thirty-eight have died and twenty-seven have
improved. Amongst the latter, only those are included
who have decidedly increased in weight, as well as showing
improvement in other respects. The precautions against
infection and against the dissemination of dust are said to

be very complete. No handkerchiefs are allowed, but linen rags are used instead, and a large number of spitcups. The patients are clad in aseptic dresses; they disinfect their buccal cavity regularly and thoroughly, and great care is exercised to keep the whole body exceptionally clean. Excepting on visiting day, little or no dust is brought in from outside. The windows are kept open day and night; and twenty-two reclining chairs with pillows of oats have been provided in the garden for rest in the open air in tents. A very copious diet is adopted. Beyond subcutaneous injections of guaiacol in 1000 parts of sterilised oil, the drugs usually administered elsewhere have been employed, arsenic, quinine and tannin being largely prescribed.

The arrangements at the Lariboisière are said to be less satisfactory, owing to the old-fashioned wards, which have been partially sub-divided into compartments. Dr. Unterberger was probably the first to introduce the treatment of consumptives in hospital sanatoria (see p. 281). In England, owing to the existence of special hospitals, the problem has been differently solved.

CHAPTER XXV.

Since the late Dr. Brehmer founded his sanatorium at Görbersdorf in 1859, the number of these institutions has been steadily increasing in Germany, so that there are now a dozen or more for paying patients, besides open health resorts where a similar treatment is carried out, and a large and increasing number of closed sanatoria for the poorer classes. The latter are described in chapters xxxii. to xl. As might be expected, most of the sanatoria for paying patients are in or near the mountains. There are several at Görbersdorf near the Riesengebirge; one is at Reiboldsgrün in the Erzgebirge; several more (Altenbrak, St. Andreasberg, Sülzhayn) in the Harz; two at Rehburg in hilly ground near Hanover; two at Hohenhonnef and Laubbach respectively, on the Rhine; the Falkenstein Sanatorium in the Taunus Mountains; while others are in the Black Forest (Nordrach, St. Blasien, Schömberg, and Marzell).

Consumptive patients are also treated by hygienic methods at Ems, Brückenau, Sophienbad, Kissingen, Baden Baden, Blankenhain, Dillenburg, and other health resorts. There are however no closed sanatoria in these places; and most of them suffer from the inevitable drawbacks attaching to open sanatoria in fashionable health resorts.

(148)

GERMAN SANATORIA FOR PAYING PATIENTS.

		Feet.	Beds.
Brehmer's	Silesia	1840	156
,, 2nd cl. . .	,,		179
Römpler's	,,	1805	120
Weicker's	,,	1837	30
Reiboldsgrün . . .	Saxony	2296	108
Lahmann's . . .	,,	780	———
Altenbrak	Harz	1017	? 20
St. Andreasberg, Jacubasch	,,	? 2160	17 +
St. Andreasberg, Ladendorf	,,	,,	12 +
Sülzhayn Fernsicht . .	,,	? 1640	10
,, Village . .	,,	? 980	9 +
Rehburg, Michaelis . .	Hanover	800	20
,, Lehrecke . .	,,	? 800	22
Hohenhonnef . . .	Rhine	774	109
Laubbach	,,	260	113
Falkenstein . . .	Taunus	1312	112 +
Nordrach	Black Forest	1470	45
St. Blasien . . .	,,	2625	62
Marzell	,,	1378	?
Schömberg . . .	,,	2130	50

CHAPTER XXVI.

THIS chapter comprises the Brehmer sanatorium with its second-class section, Dr. Römpler's sanatorium, and Dr. Weicker's private sanatorium. The "Krankenheim" is described in chapter xxxiii.

THE BREHMER SANATORIUM,

which is the oldest in existence, is situated at Görbersdorf, a small village in the valley of the Steine, between the Eulengebirge and the Riesengebirge, in Upper Silesia, near the Bohemian frontier. The valley, which runs from north-west to south-east, is sheltered by densely wooded heights, which reach an altitude of 800 to 900 metres above the sea. It is here that Dr. Hermann Brehmer started his treatment in 1854, and from 1859 onwards erected the picturesque establishment which bears his name, the right wing being built in 1862, and the left wing sixteen years later. Brehmer died in 1889, after which the sanatorium was successively directed by Dr. Felix Wolff-Immermann, Dr. Achtermann, and Prof. Dr. Rudolf Kobert formerly of Dorpat University, who manages it for the benefit of Dr. Brehmer's heirs.

The main climatic features of Görbersdorf are its atmospheric purity and freedom from dust, dryness of soil, shelter against strong wind, and cool summer temperature with abundant sunshine. Its altitude is insufficient to give it the character of an alpine climate, although the barometric pressure is somewhat reduced.

(150)

FIG. 14.—THE BREHMER SANATORIUM, GÖRBERSDORF.

[Face page 151.

The sanatorium, which stands on porphyritic soil, 561 metres (1840 ft.) above the sea-level, has a park of 300 acres, consisting mainly of woodland, more than two-thirds of which is laid out in over nine miles of walks. Some of these are on level ground, while others are inclined at various gradients leading uphill from the sanatorium ; and all of them well provided with seats and summer-houses, so that the patient can, under medical advice, take exercise proportioned to his strength, and return downhill when he is getting tired. For robuster patients there are also walks up the mountain side to a height of about 200 metres (656 ft.) above the building.

The sanatorium itself consists of a main building (fig. 14) with a number of isolated villas. The chief part of the structure is a huge Gothic pile, with towers, turrets and arches, consisting of three buildings of different dates, united in one line by covered passages and glazed galleries forming the winter garden. The central portion contains the doctor's quarters and administration. The right wing, a square building with a square tower attached to it, goes by the name of " Old Curhaus " ; the left being called the " New Curhaus ". On the ground floor are the two dining saloons, library, ladies and gentlemen's conversation rooms, conservatory (or " cold winter garden ") and palm house (or " warm winter garden ") with a spring of water. The new Curhaus has a handsome entrance and staircase, ornamented with frescoes and useful maxims. There are in the old and new Curhaus eighty-eight bedrooms with 104 beds. Those on the first floor of the left wing open on to an arcade with Gothic arches ; those on the two upper floors have mansard windows. The walls are oil-painted up to a height of 1½ metres. The floors are also painted or covered with linoleum. No large carpets are permitted ; but stuffed chairs, rugs and curtains are not objected to. The ventilation is partly by open windows, partly by air shafts for the foul air. The dining-rooms have " Kosmos ventilators," which renew the air five times per hour. The heating is partly by hot air, partly by hot water, partly by tiled stoves for burning wood. The rooms are lighted by oil lamps and candles. There is a lift for patients. There were in Dr. Brehmer's time no "liegehallen" or open-air galleries, as he objected to them, and laid more stress on exercise in the open air than on the rest cure. Under Dr. Kobert, however, several roomy fresh-air galleries have been built.

Of the three villas, the "White House" is close to the new Curhaus, and contains twelve bedrooms with thirteen beds; the "New House," with eighteen rooms and twenty-two beds, is a little farther off; and the "Villa Rosa," with sixteen rooms (one double-bedded), is on a higher level in the woods. The establishment also includes chemical and bacteriological laboratories, a meteorological observatory, a large library with several thousand volumes (medical and general), a model dairy with sixty cows under strict veterinary control, stables, laundry and disinfection apparatus. There are two hydropathic installations—one in the main building, one in a cottage in the park. The sewerage is dealt with according to the "Heidelberg Tonnen System"; the disinfection of the waste waters according to Dr. Hulwa's system.

The patients, who are exclusively consumptives, are treated much as in other establishments. In febrile stages they rest in the summer-houses, on the balcony, or in their own rooms with open windows. In other stages they walk out in the park. There are five *meals* per day, about 7, 10, 12·30, 4 and 7 o'clock, the time being notified by horn signals. Patients are woke at 6·45, and take a walk after breakfast. They rest half an hour before midday dinner, and go for a walk after it; rest again from 6 till 7 P.M.: after dinner they stay in the park in summer, in the assembly rooms in winter, and go to bed at 9 P.M. All the patients take a large quantity of milk; those with gastric disturbance have kefir from the farm. Febrile patients are given a glass of milk every hour. The *hydrotherapy,* which is applied in the morning, is of all degrees from washing with acidulated water to the douche. Dettweiler's flasks and spittoons filled with sawdust are used for the *sputa*, which are mixed with lime and buried, or else mixed with sawdust or peat mould and burned. There is a Protestant chapel near the sanatorium, and a Roman Catholic church at Friedland, from which a priest comes over once a fortnight. For the patients' amusement a series of concerts, theatricals and the like are arranged.

The charges are from 56 to 87 marks per week in summer, according to rooms, 7½ marks more in winter. In addition there is an entrance fee of 25 marks, and extra for beverages, baths, rubbings, douches, inhalations, disinfection (5 marks), bedding, and a few other things. Patients buy their own spitting flask and thermometer. Visitors and friends, attendants, and children under ten, are received at reduced rates.

Statistics.—The average number of patients received into the sanatorium of late years has been 445 per annum, with a total of 36,348 days of treatment, showing a steady increase in popularity.

Mode of Access.—The nearest stations are Friedland, on the line from Breslau to Freiburg; and Dittersbach, on the Silesian Mountain Railway. The former, which is 6 kilom. (nearly 4 miles) from Görbersdorf, is best suited to those coming from Breslau or Austria ; the latter, 9 kilom. (5½ miles), is more convenient for the majority of patients. Carriages should be ordered from the sanatorium. Görbersdorf is a post and telegraph station, but with no night service.

Second-Class Patients.

Since April, 1894, patients have also been received under treatment at Görbersdorf at reduced rates. For this purpose a number of houses near the sanatorium, with 149 bedrooms and 153 beds, have been utilised, in addition to twenty-six bedrooms let to the institution in the village. Each

house has a common assembly room, while kitchen, dining-
room, music and reading rooms are in a separate building.
The rooms are described by Hohe as roomy, lofty, light and
cheerful. Patients have the use of the park and woods of
the sanatorium. The dietary is about the same as in the
other establishment, but somewhat simpler; the treatment
in other respects being identical.

The charges are from 32 to 40 marks per week according
to room; the entrance fee is 15 marks. A reduction is made
for friends, visitors, and children under seven. In winter
4 marks extra per week are charged for heating and light.
Wine and drugs are also extra. A reduction is made in
railway fares for the patients of this section, as in the case
of other sanatoria for the less wealthy classes. Dr. Tir-
mann, second medical officer under Dr. Kobert, is in charge
of the second-class patients.

The Brehmer sanatorium is by far the largest in existence. Including first and second class patients, a total of
335 patients can be accommodated, twenty-six in the village
houses with which arrangements have been made, the rest
in the various buildings belonging to the sanatorium. In
Dr. Brehmer's time the largest number under treatment at
one time was about 330; of whom two-thirds had to be
lodged in the houses of the village. After his death there
was a diminution in the popularity of the sanatorium; but
under Dr. Kobert the number of its patients has again increased, so that there were this year for a time as many as
315 patients simultaneously under treatment. There are
now eight medical officers, three bath attendants and six
nurses attached to the institution. Statistics of this sanatorium may be found at p. 53.

The sanatorium has been somewhat severely criticised by
Léon Petit and others, mainly on the ground that hygienic
requirements have been sacrificed to architectural beauty.
Doubtless much money has been spent on unnecessary
adornments; and the heating, lighting, and ventilation
were open to improvement; but there is reason to believe

FIG. 15.—Dr. Römpler's Sanatorium, Görbersdorf.

[Face page 155.

that under Dr. Kobert's energetic management many re-
forms have been effected.   The sanatorium grounds are
probably the finest in existence ; and the place will always
be of special interest to practical physicians as the first in
which the hygienic methods of treatment of consumption
were systematically carried out, and the value of " closed
sanatoria " demonstrated.

DR. RÖMPLER'S SANATORIUM.

This establishment, which is placed within a few hundred
yards of Brehmer's establishment at Görbersdorf, was
opened in 1875 by Dr. Römpler, in two houses belonging to
Freiherr v. Rössing, which were shortly after bought by the
doctor.   The main building of the sanatorium, which is
550 metres (1805 ft.) above the sea-level, is in a strip of
land of 25 to 30 acres, separated from the rest of its
grounds in most of its extent by those of the Brehmer
establishment.   The total area which belongs to it is about
230 acres, 70 being woodland, and is laid out in the same
style as that of the larger establishment, but has not yet
had time to attain to the beauty of the latter.   However, it
has every requirement for successful treatment, and the
sanatorium is stated to be comfortable and well managed.
The park is on the slopes of the mountain side, and rises to
a height of 800 metres above the sea-level, a Swiss châlet
with balconies and verandahs, which is warmed in winter,
being placed in the middle of the grounds for patients to rest
in, and get books and refreshments.   Other resting places
and summer houses are scattered through the park.

The main building, or Curhaus (fig. 15), consists of two
parts united by a large and well-ventilated winter garden.
Each part consists of a centre and two projecting sides,
four storeys high.   The main entrance opens on the one
hand into the winter garden, on the other into a large
dining saloon capable of seating 200 persons.   Near this
are a reading room, and waiting and consulting rooms for
the medical staff.   On the first floor are other assembly

rooms, music, billiard and ladies' rooms. The Curhaus has over 100 bedrooms for patients, all of them comfortably and suitably furnished, and some provided with balconies. A large fresh-air verandah surrounds the Curhaus from north-east to south-west Communicating with this is a long covered walk extending into the grounds and provided with a concert hall. There is also a large terrace with glass roof on the first floor on the south-west side of the Curhaus, affording a fine view of the mountains. The sanatorium is so constructed as to facilitate cleansing. The walls in the corridors and bath-rooms are oil-painted; elsewhere of washable paper. The floors in the bath-rooms, winter garden, reading-room, balconies and terraces are tiled; in the bedrooms of lacquered wood. The sanatorium is heated by Perkins' system of hot-water pipes; heating and ventilation of baths and douche rooms being effected by hot-water pipes from the kitchen. Chimneys are only present in one of the common rooms. The water supply is from a spring on the borders of the park 70 metres above the sanatorium, and is therefore able to supply every floor. There are baths in the Curhaus, and baths and douches in a separate bath-house in the park. The water-closets in the Curhaus are provided with a good flush, the waste waters of every kind being filtered through coke one metre below the surface in the park, and the resulting effluent being pure and clear. Two separate villas in the park near the Swiss châlet are devoted respectively to convalescents and to those who wish to live apart from the other patients. There is also a chapel in the park, where service is held once a fortnight. A well-stocked library and numerous games are also provided.

The treatment is on the usual lines, none but consumptives being admitted. *Sputa* are received into Dettweiler's flasks or into spitting cups containing water. Linen and clothing generally are disinfected by a steam disinfector. Whenever a room is vacated it is disinfected by Schering's formalin disinfector. The patients take rest or exercise according to their medical condition. There are two trained nurses in the establishment, in addition to three male and two female rubbers.   Cod-liver oil is occasionally given in winter ; more frequently carbonate of guaiacol. Camphorated oil is used for external applications.   The results with tuberculin have not been encouraging, so that it is only used where patients urgently desire it.   Dr. Römpler lays stress on the necessity of individualising in treatment of patients.   Five *meals* a day are provided : a first and second breakfast, mid-day dinner of soup and three courses, afternoon tea and hot or cold supper.

The charges are, for the room, 7 to 25 mks. per week according to position ; for board and attendance, 30 mks. ; doctor's fees, 7 to 10 mks. ; entrance fee, 21 mks., or 30 for a family.   This entitles to the use of newspapers and periodicals, the use of the park and its special arrangements, concerts, etc.   In winter 6½ mks. per week extra are charged for heating and lighting.   Other extras are wine, drugs, the use of bedding (1½ mks.), extra milk and baths.   There is no extra charge for special diet ordered by the doctor.   Douches are provided free.   Children under ten are received at half-price, and pay no entrance fee. For servants and attendants a reduction is also made.

There is accommodation for 120 patients in the sanatorium and its annexes.   For statistics see p. 53.

Both Léon Petit and Hohe speak highly of the arrangements in this sanatorium.   The nearest railway stations are Friedland, which is twenty minutes' drive from the sanatorium, and Dittersbach, which is one and a half hours by carriage.   This may be ordered from the sanatorium. For soil and climate see the Brehmer sanatorium.

DR. WEICKER'S SANATORIUM,

or the "Sanatorium of the Countess Pückler," owes its origin to a convalescent home for poor people founded by the late Countess Marie v. Pückler in 1883, at Schmidtsdorf, less than a mile to the west of Görbersdorf, and in the same valley.   The institution, which belonged to the village of Schmidtsdorf, was acquired in 1893 by Dr. Hans Weicker, who converted it into a sanatorium for consumptives, and those suffering from diseases of the nose, throat and larynx. It is a small establishment, and its owner intends it to remain so, believing that in this way it is possible to give more individual attention to each patient. The sanatorium is sheltered by mountains to the north, and at a greater distance by wooded spurs to the east and west. The woodland belonging to the establishment amounts to scarcely 6½ hectares (16 acres), but the patients have access to the adjoining woods of Count Pless, according to an agreement with the owner; so that plenty of sheltered walks are available, with resting places and pavilions and beautiful views.

The sanatorium (fig. 16) stands at 560 metres (1837 ft.) above the sea-level. It is an oblong block four storeys high, with a receding centre portion in front of which is a terrace, and higher up two deep balconies. The lower balcony is continued round the house to the east and west ends. There is a small open-air shelter or "liegehalle" touching the eastern end of the house. The rooms are simply furnished; the walls being partly oil-painted, partly colour-washed, partly papered, and the floors painted or covered with linoleum. Every room has a chimney, and is heated by closed stoves. The lighting is at present by petroleum, but electric lighting is to be introduced. The closets are turf-mould closets, the waste water being clarified by filtration.

The treatment in the main resembles that in other German sanatoria. Patients take exercise in the grounds and woods, excepting before dinner and after meals. Those who are febrile or dyspeptic remain at rest on couches in the open air. Dr. Weicker has introduced some modifications in the couches used, to enable various horizontal or reclining positions to be assumed. A small carriage is used to take slightly febrile patients across the park. Systematic respiratory exercises are prescribed in suitable cases.

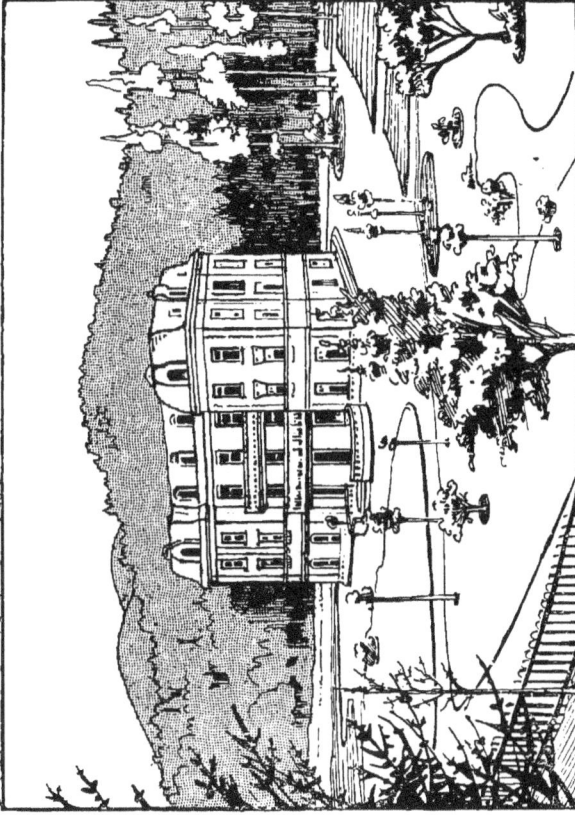

FIG. 16.—DR. WEICKER'S PRIVATE SANATORIUM, SCHMIDTSDORF, NEAR GÖRBERSDORF.

[Face page 158.

Hydrotherapy is employed, but not in the form of douche. In the morning the patients are rubbed with a wet glove with water at 93° F., and replaced in a warm bed without being dried. For night sweats, sponging with solutions of spirit or camphorated spirit is used. Another favourite application is a hip bath at 93° F. followed by affusion at 86°. General massage is also applied every other day if the appetite is unsatisfactory. Moist compresses are used at night to counteract congestion of the lungs. Drugs are but little used, and mainly to combat urgent or troublesome symptoms. Cod-liver oil is seldom used ; very little alcohol is given, and only a moderate amount of milk. So-called specifics are not used. Daremberg however states that creosote and guaiacol are often prescribed. The physician lives in the house, and takes his *meals* with the patients. Milk or soup is served in bed before the patients rise. Then follows a breakfast of coffee or cocoa, kefir or soup with bread and butter, biscuits, etc., after this a light lunch of milk with bread and butter, followed by mid-day dinner of soup, fish, meat, vegetables and stewed fruit or pudding. In the afternoon another meal is given like the breakfast. The last meal is a hot or cold supper. No extra charge is made for special diet ordered by the physician. There are two trained nurses in the house. *Sputa* are disinfected with lysol, the linen by boiling, the rooms with formalin vapour.

The charges in summer amount to from 41 to 52 mks. per week ; in winter (1st October to 15th May), 3½ mks. extra are charged for heating and lighting. There is an entrance fee of 10 mks., or 15 for families. A charge of 5 mks. is made for the first medical examination. There are a few extras. The establishment will accommodate twenty-five to thirty patients. In 1896 it received fifty-nine patients, with 3802 days of treatment.

For further particulars, see the Brehmer sanatorium. Dr. Weicker is also the medical director of a sanatorium for poorer patients, described at p. 209.

CHAPTER XXVII.

FOR the sake of convenience, the Saxon sanatoria of the Harz district are described in the next chapter. The only other sanatorium which, strictly speaking, belongs to this group is the one at Reiboldsgrün. A short description of Dr. Lahmann's sanatorium near Dresden is, however, added, although it is not specially an institution for consumptives.

DR. DRIVER'S SANATORIUM, AT REIBOLDSGRÜN,

is situated on the lower slopes of the Erzgebirge in Saxony. These mountains, which form the natural frontier between Saxony and Bohemia, and reach an altitude of 750 to 800 metres (2460 to 2624 ft.), with a few peaks of 1200 metres (3937 ft.), consist of a number of parallel chains separated by valleys, each containing a tributary of the Elbe. Here, in the little valley of the Zinsbach, in the midst of a huge forest of pine trees, over ninety miles long by twelve to eighteen wide, is situated the sanatorium of Dr. Driver, at an altitude of 700 metres (2296 ft.) above the sea-level.

The district has been known and frequented since 1725 for its ferruginous waters, and had a little thermal establishment, which was acquired in 1873 by Dr. Driver, and converted into a sanatorium for consumptives, after the model of Brehmer's at Görbersdorf. Dr. Driver still owns the establishment, and lives on the spot, but has entrusted the management since 1892 to Dr. Felix Wolff-Immermann, formerly at the head of Brehmer's sanatorium.

(160)

The place is on volcanic soil, permeable, yet free from dust. The climate is a sheltered hill climate, with quiet air and little variation in temperature. The sanatorium stands in a park of seventy-six and a half hectares (about 189 acres), provided with numerous walks, resting-places and summer-houses, and very diversified in character. It also contains an artificial lake, which is in winter used for skating and sledging. The situation is a most convenient one for a closed sanatorium, owing to its isolation, as there is no village or health resort for miles, and pine woods spread on every side.

The sanatorium (fig. 17) consists of a group of eight large buildings, arranged for the most part along the northern and eastern sides of a large garden, and connected by covered corridors. There are also a few more small buildings for the staff and management.

The central block, or Curhaus, which was erected in 1890, contains a dining saloon for over 100 persons, several assembly rooms (reading, music, and billiard rooms, etc.), and the kitchen department, and has a verandah along the southern side. Attached to the Curhaus is another, of which the ground floor contains the postal department and the office ; the two upper floors containing fifteen rooms, mostly to the south. It is in this building that the medical director lives. The Curhaus also communicates on the other side, by a covered corridor, with the " Villa Winterheim," which is the largest of the group. It contains on the ground floor the douche and bath rooms, drug room, waiting and consulting rooms, and laboratory. The three upper floors contain thirty-four patients' bedrooms, none of which has less than 100 cubic metres (3530 cub. ft.) air space. There is a high tower at the eastern end, and next to it a glass-covered verandah communicating with four large galleries, and through them with the " Villa Wiesenhaus ". This, like the preceding, is very comfortably furnished ; it has twenty-six patients' rooms on two floors. A little lower down is " Hugo's Ruhe," with fifteen rooms, some of which

11

are occupied by the assistant medical officer. Near the Curhaus is the "Turmhaus," containing on the ground floor the administrative department; on the upper floors twenty-two rooms, mostly smaller than the others. Beside this is the "Villa Karlsruhe," with twelve good rooms; and Dr. Driver's house, "Villa Mathildenruhe". In addition to these villas, there are stables and farm buildings, laundry with steam disinfector, but no private dairy. About a quarter of an hour's walk up-hill in the park is a house (also under medical control) where patients can obtain rest and refreshment, while they enjoy an extensive view over the surrounding country. Convalescents and friends of patients are also quartered here. The sanatorium altogether contains 111 patients' rooms, and will accommodate 105 to 108. Most of the rooms are 4 metres high; the smallest covers 15 square metres (161 sq. ft.), and most are considerably larger. The walls are painted with oil colours for 1½ metres, and lime- or colour-washed above this. The floors are mostly covered with linoleum. The Curhaus and "Villa Winterheim" are heated by low-pressure steam: the other villas, partly by cast-iron stoves, partly by tiled stoves. The lighting is by petroleum lamps, the "regenerative lamp" being used in the larger rooms. There are no special ventilating contrivances. There is considerable difference in the appointments of different parts of the establishment; but the furniture is generally simple and readily cleansable. The establishment has its own spring of water, and, in addition, a steam pump in case of drought.

The sanatorium is exclusively for consumptive patients, and has been strictly so since early in 1892. Only those are admitted who are likely to benefit from the climate. If there are no signs of improvement within the first fortnight, they are advised to leave the establishment, as, in Dr. Wolff-Immermann's experience, this is a reliable test of the suitability of the climate. Out of 366 patients who left in 1897, 11 per cent. did so before the twentieth day, and half this number before the eleventh day. The average duration of stay is sixty-six days. Doubtless in many cases patients leave too soon; but some of those who benefit at first become febrile again later on, and yet promptly recover their health on removal to a lower altitude. This illustrates the fact that there must be a certain relation between the stimulating qualities of each health resort and the reactive powers of the patient. The number of those who apply for admission at Reiboldsgrün is steadily increasing. In 1892 there were 264 treated; in 1897, 377 were admitted. The average for the five years ending in 1896 was 311 patients, with 21,874 days of treatment; in the previous period, 270 patients, with 19,833 days of treatment. By far the largest number were Germans, this being a much less cosmopolitan sanatorium than most of the other large establishments in Germany. There are on an average seventy in the sanatorium at the same time.

The treatment is on the usual lines. Febrile patients rest, while others take exercise in the open air, spending about ten hours out of doors every day. It has been increasingly the custom to make use of the shelters in the woods for rest, instead of the verandahs. The occupation of patients is very carefully controlled. There is a good library; some of the more exciting books are forbidden to febrile patients. Musical instruments and unobjectionable games are also provided. There are periodical concerts and entertainments. In winter, those who are fit for it go sledging, or even skating. There are five *meals* a day, as at other German sanatoria. The medical director is not a believer in the excessive administration of milk, or in unnecessary "stuffing".

Sputa are received into spittoons containing water, and are emptied once or twice a day into the water-closet. The linen is disinfected by steam. As regards drugs, the chalybeate spring in the grounds is given when climatic treatment alone does not give satisfactory results; in other cases, symptomatic treatment is adopted, very little else being given excepting antisyphilitic remedies where called for, or quinine and arsenic where there is a malarial taint.

Of those treated, 80 per cent. were "better" when they left. Two died, one from tubercular meningitis, the other from severe hæmoptysis, which came on while he was improving in health. To appreciate the value of the treatment, Dr. Wolff-Immermann's careful report and analysis should be studied.

There is one assistant physician in summer, two in winter. One trained nurse is found to be sufficient.

The charges are 45 to 63 marks per week, with an entrance fee of 10 marks, or 20 for a family, and a few extras. Reduction is made for children, visitors, and attendants. Since the opening of the Albertsberg Sanatorium no "second-class" patients connected with insurance societies are received.

Evangelical service is held once a fortnight in summer, once a month in winter. A chapel is, however, to be built for the common use of the two sanatoria. Roman Catholics can find a church at Auerbach.

The nearest railway station is at Rautenkranz, nearly three miles off, on the line from Chemnitz to Aue and Adorf. This is best for patients coming from Bohemia and S.E. Saxony. For others, Auerbach station (about 4½ miles), on the line from Berlin and Leipzig to Hof and Munich, is more convenient. Carriages should be ordered from the sanatorium.

Dr. Lahmann's Sanatorium,

at Weisser Hirsch, near Dresden, although not a special sanatorium for consumptives, is here alluded to, inasmuch as it receives cases of early phthisis amongst others, and the methods of treatment are to some extent similar to those of the sanatoria for consumptives.

Many of the ailments which flesh is heir to are directly or indirectly caused by unnatural and unhealthy conditions of life. People live in sunless and ill-ventilated rooms,

Fig. 18.—Dr. Lahmann's Sanatorium.   Sleeping Box.

[*Face page* 165.

wear unsuitable clothing, take too little exercise or ill-con-
sidered spells of sedentary work, or an unsuitable dietary,
or give themselves up to various forms of self-indulgence.
The evil results of such hygienic errors can to a great extent be remedied by hygienic means; and this is what Dr.
Lahmann's sanatorium professes to do. Only "natural
remedies" are employed, such as water, light, fresh air,
massage and regulated exercise, dietetics, electricity, and
in certain cases mental suggestion. Drugs of every kind
are dispensed with, in order to avoid the possible drawbacks
incidental to their use.

The sanatorium is situated on a plateau 238 metres (780
ft.) above the sea-level, in the health resort Weisser Hirsch,
a village of about 1200 inhabitants, which in summer
receives the overflow of visitors from the sanatorium.
Sheltering the village to the north and north-east is the
extensive forest called the Dresdner Heide, which is provided by the local Improvement Society with good paths,
benches, shelters, and the like. The extensive grounds of
the sanatorium, which also include a portion of woodland,
are contiguous with the public woods.

The sanatorium consists of a main building containing
various reception rooms and dining saloons, a bathing
establishment and gymnasium, and a number of separate
villa residences. The reception rooms in the main building
are lighted with electricity, and communicate with the
bathing establishment by means of a covered walk. In
addition to these buildings, there are a number of open
shelters for the fresh-air treatment, fitted up for sleeping
as well as for use in the day time (fig. 18). Another
feature of the place consists in the systematic air-baths in
specially-arranged and sheltered spots, designed to restore
the lost functions of the skin, and sun-baths to stimulate
nutritive processes.

Hydrotherapy is extensively used. The ordinary dietary
consists of five *meals*, with a smaller proportion of meat
than is usual elsewhere in Germany. In certain cases a

vegetarian diet with eggs and milk is prescribed. Special
attention is paid to the proportion of various salts in the
food and in the excretions, as determined in the chemical
laboratory. The buildings are heated by means of central
steam apparatus. The sewerage has been perfected at heavy
cost, and the water supply is said to be excellent and abundant. The sanatorium has a private dairy farm of tested
cows. For the amusement of the visitors, in addition to
books and papers, concerts, musical evenings, dances in
summer, bowls, croquet, billiards, cycling, and pony sledging in winter, are provided.

The duration of treatment varies according to the nature
of the case—in constitutional ailments at least six weeks is
advised; and in long-standing maladies as many months as
there have been years of illness. There are six doctors in
residence in addition to Dr. Lahmann, including one lady
doctor. During 1897 over 2000 visitors (or, including the
village, over 3000) were received, of whom 221 came as
attendants or for change of air. There are on an average
300 visitors at one time in the sanatorium in summer, and
100 in winter, of the most various nationalities.

The charges, including board, lodging, and medical treatment, are from 70 to 91 marks per week, with a few cheaper
rooms for those without means. Reductions are made in
the case of children, and where two sleep in one room.
Those who lodge in the village pay 62 marks, or, for board
or treatment alone, 42 marks per week.

The sanatorium is an hour's drive from the station, which
is connected with Loschwitz by a wire-rope railway, and
thence with Dresden (Neustadt) by two electric railways
and a steamboat service. The telegraph and telephone
office is opposite the sanatorium.

CHAPTER XXVIII.

No important closed sanatoria for paying patients exist in this beautiful district of Germany, although there are several for the less wealthy classes. A small sanatorium exists at Altenbrak ; another has been built at Sülzhayn, which, however, is being occupied by working-class patients until the completion of the large sanatorium destined for their use. Drs. Ladendorf and Jacubasch treat patients at St. Andreasberg by the hygienic open-air method, but their establishments are open sanatoria, not exclusively for consumptives. The " Felixstift " of Dr. Ladendorf, however, is a closed sanatorium for the less wealthy middle-class patients, resembling in this respect the "second class" section of the Brehmer sanatorium. It is described at p. 231.

THE ALTENBRAK SANATORIUM

is in the " Bodenthal," which many regard as the most beautiful part of the Harz district. It lies in a valley open to the south and protected by mountains covered with pine woods, which are only a few steps from the sanatorium. It is 310 metres (1017 ft.) above the sea-level, and has accommodation for twenty patients, unless the contemplated enlargement has already taken place. It is managed by an assistant of Dr. Pintschovius, who is himself a resident in the neighbouring village of Ketzin, and in charge of an open sanatorium for the poorer classes. The charges are 42½ to 46 marks per week ; an entrance fee of

10 marks is required, or 15 for a family; and 5 to 10 marks are charged for the first consultation. Children are received at reduced rates. The nearest railway stations, which are one to two hours by carriage from the sanatorium, are Blankenburg and Thale. I have been unable to obtain full particulars concerning this sanatorium.

St. Andreasberg

is a small town of nearly 4000 inhabitants, with a visiting population of nearly as many more, who visit the place chiefly for the sake of hydropathic treatment or for change of air. It is situated partly on the side of a long hill, partly in some adjacent valleys, the upper part being 660 metres (2165 ft.) above the sea-level, and the lower part some 500 ft. lower down. It is said to be the most elevated town in northern and central Germany, and has long been known as a cool summer resort. The upper part is laid out as a broad avenue bordered with trees, and is cool and bracing from its elevation and its proximity to open moor and meadow land. In the lower part the red-roofed wooden houses are more closely massed together, and the streets unusually steep though fairly wide. Mining operations are carried out in some of the neighbouring valleys, but there are no large factories to sully the purity of the air. The place has two hydropathic establishments, and is said to have 700 lodging-houses,[1] besides half a dozen good hotels, a concert hall and theatre, and a "Kurpark": electric lighting and electric trams are, it is said, to be introduced before long; but with all this, it remains a quiet and unpretentious little town, rather than an ultra-fashionable health resort. The town as a whole is sheltered from keen winds by mountains to the north and east, and by wooded hills in other directions; and from the purity of its air, the dryness of the soil (which is of shale), and the absence of dust, forms a very suitable resort for consumptives in an early

[1] R. W. Felkin, " Lauterberg and St. Andreasberg," *Proc. Med. Jour.*, July, 1892.

stage, who have of late years frequented it in winter as well as summer. The winter climate is somewhat warmer than the neighbouring lowlands. The mean temperatures for the various months are said to be as follows :—

January,	.	.	.	− 1·2°	July,	.	.	.	.	+ 15·0°
February,	.	.	.	− 1·3°	August,	.	.	.	+ 13·7°
March,	.	.	.	− 0·5°	September,	.	.	+ 11·8°
April,	.	.	.	+ 5·4°	October,	.	.	.	+ 6·7°
May,	.	.	.	+ 12·9°	November,	.	.	+ 2·1°
June,	.	.	.	+ 15·9°	December,	.	.	− 0·6°

Mean annual temperature, 6·7°; mean bar. pr., 705 mm.; mean humidity, 89 per cent.; prevailing winds, W. and S.W.

The surrounding country is exceedingly varied and picturesque, with open country in some directions, and beautiful pine-clad mountains with deeply cleft valleys in others. The water supply is exceedingly abundant, and is said to remain so even in seasons of drought. As in most foreign health resorts, visitors have to pay a *kurtaxe.*

The two hydropathic establishments—which belong respectively to Dr. Ladendorf and Dr. Jacubasch—are both in the middle of the town, in situations which are rather sheltered than bracing. Dr. Ladendorf receives a few consumptive and other patients into his own house for treatment; others are quartered in neighbouring houses. It is scarcely to be called a sanatorium for consumptives. Dr. Jacubasch's sanatorium, which is close to his bathing establishment, consists of a pleasant-looking villa, which is provided with wide covered verandahs and balconies, and has accommodation for ten first-class patients. Arrangements have also been made with a neighbouring lodging-house for the reception of seven second-class patients. Surrounding these various buildings, and the doctor's own residence, is a large garden; and adjoining this is the " Kurpark," with its sheltered walks and summer-houses, and its artificial lake and fountain. Dr. Jacubasch does not accept any advanced cases for treatment; nor does he confine himself to consumptives. These are treated according to the ordinary methods; but special stress is laid on the utility of pine baths and inhalations.

The charges are from $4\frac{1}{2}$ to 6 marks per day, including board, lodging, attendance, and cleaning of clothes and apartments. Wine, beer, milk, baths, heating and lighting are extras. The meals include a first and second breakfast, dinner of four courses, afternoon coffee, and supper or high. tea. For medical attendance, 10 to 20 marks are charged per week, and 5 to 10 marks for the first consultation. Second-class patients pay 4 marks per day, or, where two sleep together, $3\frac{1}{2}$ marks. Reductions are also made in the other charges. If any patients are taken seriously ill, there are four trained nurses in the town, whose services may be obtained.

Dr. Jacubasch has published statistics of 500 consumptives treated in his sanatorium.[1] More recent figures,. including 470 additional cases, are as follows: Greatly improved, 90·4 per cent.; better, 4·8 per cent.; no change, 2·1 per cent.; died, 2·7 per cent.

St. Andreasberg is the terminus of a small branch railway from Scharzfeld, on the line from Nordhausen to Northeim. It can be reached in about $4\frac{1}{2}$ hours from Hanover.

THE SÜLZHAYN SANATORIUM,

for paying patients, called "Fernsicht," is described at p.. 238, together with the adjoining establishment for the poor; the " Felixstift " and the Oderberg Sanatorium at pp.. 231, 226.

---

[1] *Lungenschwindsucht und Höhenklima*, Encke, Stuttgart, 1887 ; *Prager Med. Woch.*, 1892, No. 29 ; *Deut. Med. Woch.*, 1889, No. 27.

CHAPTER XXIX.

IN travelling southwards from Hamburg or Bremen, through
the North German plain, a range of hills is met with to the
west of Hanover, which at once attracts attention by its
beautifully wooded and diversified scenery. It is about six
miles long, with a direction from south-east to north-west,
and rises in parts to a height of nearly 500 feet above
the sea-level. Near it, to the north-east, is the Steinhuder
lake ; while the city of Hanover is some 25 miles to the
east. The ridge is part of an ear-shaped patch of wealden
strata, isolated in the alluvial plain. Its western slope is
gentle, and is covered with conifers, oaks, and other trees ;
on the eastern slope, which is more abrupt, beeches pre-
dominate. It is in a deep notch of this eastern side that
the village of Bad Rehburg is situated, well known for
its mineral springs and bathing establishments, and for
its "whey cure" in summer. The bathing establishment
and much of the surrounding country are in the hands of
government. The village consists of a few wide streets
of unpretentious houses, and of two main avenues at right
angles to one another, overshadowed by beautiful trees,
and lined with scattered country villas and rural hotels,
each with its own patch of garden. The climate is a
warm and moderately moist valley climate, chiefly remarkable for its freshness in summer, and for the absence
of strong wind. The chief visiting season has for many
years been from May to mid-September; but more recently the place began to be visited in winter also by
consumptive patients ; and when Brehmer's example led to

the erection of so many sanatoria in Germany for the fresh-air treatment, three such establishments were opened at Rehburg, one being for the poor of Bremen, the two others for more wealthy patients. Exaggerated notions of the infectiousness of consumption have unfortunately injured the popularity of Bad Rehburg, which has consequently fallen into undeserved neglect. Those who wish for a quiet rural retreat, with beautiful country in the immediate neighbourhood, would do well to remember this once fashionable health resort. It is served by a narrow-gauge railway, which runs from Wunstorf, at the junction of the line from Hanover to Bremen with that from Berlin to Cologne and Holland.

DR. MICHAELIS' SANATORIUM,

which was specially built in 1886 for the hygienic treatment of consumptives, is situated in the upper part of the village, about 300 feet above the sea-level, and is surrounded on three sides by public woods, which cover over 300 acres of ground. It stands at a distance from the road, surrounded by a garden of about three acres, remarkable for its lofty and beautiful beech trees.

The sanatorium (fig. 19) consists of a three-storey building, with a few attics in the peak of the roof, and verandahs to the south and east. It has a southerly aspect, and the rooms are so arranged as to admit sunshine into every one. The centre of the building projects in a series of open balconies and fresh-air shelters, each of which stands a little farther out than the one above. The rooms behind these are well lighted by windows in other directions. There are a few common rooms, the dining saloon having a capacity of about 1000 cubic metres (35,300 cub. ft.). The bedrooms, which are twenty in number, vary in size from 50 to 100 cubic metres (1760 to 3530 cub. ft.). They have linoleum-covered floors, whereas in the common rooms there is parquet. The walls are covered with shiny washable paper, and the furniture is also for the most part

Fig. 19.—Dr. Michaelis' Sanatorium, Rehburg, Bad Rehburg, Hannover.

[Face page 172.

readily cleansable. The heating is by low-pressure steam ; the lighting by naphthalene lamps ; but it is proposed to introduce acetylene, unless a public supply of electricity is provided. The visitors are almost exclusively consumptives, of whom only those in chronic or curable stages are admitted.

The treatment is of the usual kind. Dr. Michaelis agrees with Dettweiler in advocating rest rather than exercise in the majority of patients who come for treatment. Patients spend much time out of doors—often 12 or 13 hours a day —lying out on cushioned cane lounges in the garden, the open south verandahs, or the summer-house. There is seldom a day when they are obliged to use the closed verandah. The oleander blooms freely out of doors in most seasons; and the evenings are nearly always still and clear. In cold weather, wraps and hot bottles are provided. The *diet* is intended to resemble that of a high-class private house rather than an hotel. Plenty of milk is given, especially in small repeated doses in the afternoon. During summer patients also take whey from the royal establishment ; or, if feverish, goats' milk or kefir. The whey is said to increase the capacity for other food. In the morning it is taken hot. The goats, owing to special feeding, yield milk free from the usual strong flavour. Unlike Dr. Dettweiler, Dr. Michaelis gives very little alcohol, and never undiluted. Respiratory exercises are not favourably regarded by him, excepting for young subjects who are only inclined to consumption, and have good heart action, in which case Waldenburg's apparatus may be useful. Baths and douches are provided in the royal bathing establishment. Dr. Michaelis is a strong advocate of inhalations under a pressure of half an atmosphere, of saline solution, creosote solution, pinol and the like. For night sweats he advises a brief tepid bath, followed by rubbing with spirit or dry rubbing ; or else an infusion of freshly gathered oak twig bark. For excessive cough he gives codein ; for irregular febrile outbursts, antipyrin.

On the whole, drugs are little used in the establishment. *Sputa* are poured down the water-closets. Linen and bedding are disinfected by steam heat in the public disinfector before being used by a fresh patient. The floors and walls are regularly cleansed with a damp cloth. Dr. Michaelis lives in a separate villa in the village, an assistant medical officer sleeping in the sanatorium. Nurses are obtained as required.

The charges are from 8 to 21 marks per week for room, heating and lighting. Complete board, 5·50 marks per diem. Attendance, 1·50 marks per week. There are no extra charges for medical attendance, excepting for gynecological treatment, massage, electricity, or inhalations.

Dr. Michaelis is also the visiting physician of the Bremen Popular Sanatorium, which he attends gratuitously. This is described at p. 246.

DR. LEHRECKE'S SANATORIUM

was originally founded by Dr. P. Kaatzer, in 1886, but has been since 1896 under the care of Dr. Hans Lehrecke, who has considerably enlarged the establishment. It is not exclusively for consumptives—patients with various nasal, pulmonary, and pharyngeal affections, convalescents from other health resorts, and those who are merely in need of change of air, being also received.

The sanatorium is situated at the junction of the two main roads of the village of Bad Rehburg, in one hectare (2¼ acres) of land, close to the royal bathing establishment and Curhaus. It consists of a horseshoe-shaped two-storey building next the main road, and of detached blocks in the garden for the kitchen department, bath and douche rooms, engine-house, doctor's residence, laboratory and consulting room, etc. About twenty-two patients are received in residence. The rooms are said to be airy and light, the corridors being well lighted by the staircase skylight, as well as by windows of their own. The bedrooms are in communication with a glass-covered verandah, which is open day and night ; some also have sitting rooms attached to them. There are altogether four sheltered verandahs, three of which are over 40 feet long, and can be warmed if necessary. There is also an open sheltered terrace, where, under Dr. Lehrecke's direction, and in the charge of an attendant, the patients in summer expose themselves undressed to the air, lying on rugs or hammocks. The common rooms comprise a large dining saloon, reading room, billiard and music rooms. There is also a dark room for photography and examination with Röntgen rays. The lighting is by electricity. Heating is by closed stoves, on each of which is placed a vessel of water. There is a special water supply from a spring in the garden, of pure quality and at a temperature of 10° to 12° C. The water is filtered through sand, and pumped by the electric motor to the top of a tower for distribution throughout the establishment. Closets are of English pattern (Tornado), provided with a good flush and carefully ventilated. There is a very complete hydropathic installation in a separate wing, with brine baths, mud baths, pine baths, effervescing baths, etc., which, with massage, are applied by skilled attendants under medical direction. The rooms and corridors are laid with linoleum, and washed daily with soap and washing soda, followed by sublimate solution. When a patient leaves, his room is disinfected with formalin and repapered and repainted. Noisy and dust-producing household work is done on the flat-topped roof of a separate building, or in a large meadow adjoining the garden. *Spittoons* are provided in the corridors and rooms, partly filled with solution of sublimate and lysol. Ozone producers are used in the bedrooms and sitting rooms.

There are five *meals* a day, as well as extra milk or some substitute twice a day ; the chief meals being taken in company of the medical officers. Various alcoholic beverages are provided. Milk, which is boiled before being used, is obtained from tuberculin-tested cows and goats pastured on a field next the garden. Whey from the institution in the village and kefir are also given.

Only consumptives in the first and early second stages are admitted. They are treated by rest and exercise in the open air, and by the "air bath" already alluded to; by active and passive respiratory exercises, massage, electricity, inspiration of compressed air by Waldenburg's apparatus, followed by expiration into slightly rarefied air: methodical holding of the head, breathing through the nose, and graduated walking exercises under medical supervision. Hydrotherapy forms an important part of the treatment.

In the morning a wet rub at 24° R. (86° F.) is followed by a dry rub, and this by cold affusions or gentle douches. In the evening the patients have a tepid bath. If hæmoptysis appears, the ice compress or icebag is applied. For consolidation, wet compresses are employed. For fever, cooling baths with temperature reduced from 28° to 24° R. (95° to 86° F.), followed by ice pack. Night sweats are treated with tepid sponging in the evening, followed by sponging with spirit and vinegar. A few patients receive brine baths, mud baths, pine baths, aromatic and carbonated baths, followed by tepid douche and a rub with absolute alcohol. Dr. Lehrecke attaches great importance to the soundness of the nasal and pharyngeal mucosa, the mouth and teeth. He uses inhalations, in suitable cases, of solutions of mineral salts, atomised essential oils, ozone, and the like. Drugs are prescribed rather more freely than in some other sanatoria.

Dr. Lehrecke has two assistant physicians under him. The charges are from 56 to 68 marks per week. Extras are beverages, unusual remedies (such as tuberculin), baths and douches.

CHAPTER XXX.

Two very well-known establishments are situated in this part of Germany—Falkenstein, in the Taunus Mountains, and Hohenhonnef, in the Siebengebirge—as well as another at Laubbach, recently opened by Dr. Achtermann. Other sanatoria for poorer patients in this district are described in chap. xxxix. ; and for poor and paying patients in the Black Forest, in chaps. xl. and xxxi.

THE FALKENSTEIN SANATORIUM

was the first of its kind erected in western Germany, being opened in 1876 by a company of citizens of Frankfort, under the influence and direction of the well-known Dr. Dettweiler.

It is situated about fifteen miles north-west of Frankfort, on the southern slopes of the Taunus Mountains, at the upper end of a valley which is exceedingly well sheltered on the west, north, and east by hills of from 1500 feet to nearly 3000 feet above the sea-level. The soil consists of clay-slate, gneiss, and porphyry. The surrounding country is picturesque and well wooded, with extensive forests of beech and other trees. The climate is a moderately moist hill climate, with plenty of sunshine in summer, but a good deal of rain and fog during the cold part of the year.

The sanatorium has a south-east aspect, and stands about 400 metres (1312 ft.) above the sea, having an extensive view over the little town of Cronberg and the flat valley of the Main, which, with Frankfort on its banks, can be distinguished in the distance.

The grounds of 5·71 hectares (about 14 acres) are partly meadow land, partly garden in terraces, with clumps of bushes and trees, and are continuous on the north-west side with the extensive neighbouring public woods. Both are provided with a number of paths at various gradients, well supplied with seats, and in the grounds with summer-houses of various shapes and sizes, some of which can be turned round for shelter against wind.

The sanatorium consists of a main building with two additional villas united to it by covered corridors. The main building (fig. 20) has a central block and two wings, which diverge at obtuse angles and enclose a stone terrace with deep verandahs, under which the patients rest on cane sofas. On the ground floor of the central block are a large winter garden, a rather gloomy reading room, and cloak rooms, with a well-lighted corridor in front, looking on to the terrace and extending to the wings. Behind are the entrance, grand staircase, porter's lodge, and post and tele-graph offices. At the ends of the wings are music and conversation rooms, billiard room, and library. The hand-some dining hall, which is placed in a separate building over the kitchens, communicates through a vestibule with the eastern end of the central corridor. It is 78 ft. long, 39 ft. wide, and 32 ft. high, and can easily seat 200 guests. In front of this, diverging by an obtuser angle than the eastern wing, is an open corridor 200 ft. long, which leads to the eastern annexe. This is now entirely occupied by patients' rooms. A similar but shorter corridor on the other side of the centre leads to the western annexe, of which the basement contains the bath and douche rooms, the rest consisting of laboratories, consulting rooms, and the private rooms for the three medical officers and their establishment. The kitchen department, water-closets, and dining rooms for servants and officials are grouped around a small court near the dining saloon. The position of the

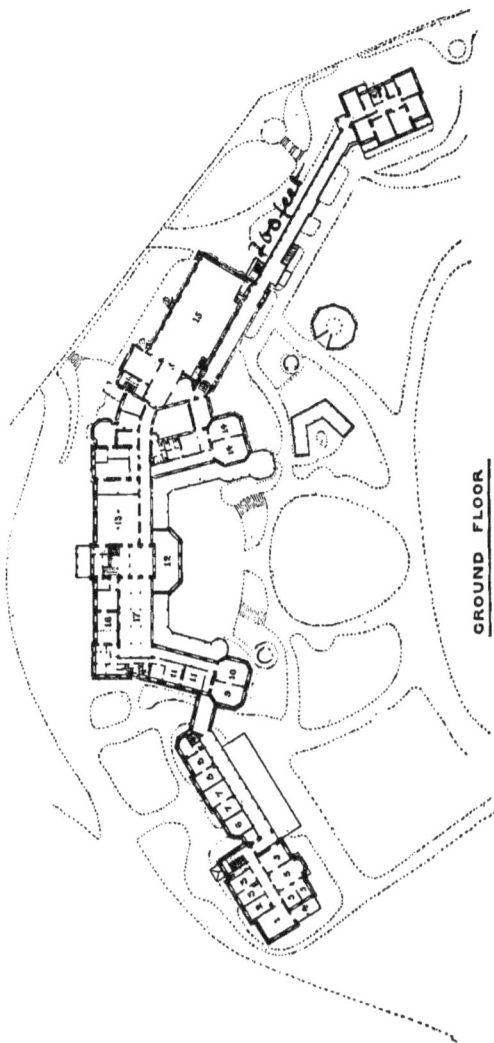

GROUND FLOOR

FIG. 20.—THE FALKENSTEIN SANATORIUM.—GROUND PLAN.

1. Board Room.
2. Visitor's Room.
3. Rooms for the Staff.
4. Verandah.
5. Consulting and Waiting Rooms and Laboratory.

6. Lavatory.
7. Mortuary.
8. Gardener's Quarters.
9. Library.
10. Billiard Room.
11. Visitor's Rooms.

12. Winter Garden.
13. Reading Room.
14. Conversation and Music Rooms.
15. Dining Saloon.
16. Office.
17. Winter Garden.

[*Face page* 178.]

latter near the garden might be turned to greater advantage by bringing the windows on the south side down to the ground, and leaving them open during meal times. At present, however, the eastern corridor, which runs in front of the dining saloon, interferes with such a desirable arrangement, owing to a difference of level. The upper part of the central block and wings, which is partially built of wood, is occupied by bedrooms on two floors, with smaller ones in the attics. Some of these bedrooms have balconies, one in the centre over the winter garden being a very large one. The bedrooms are 4 to $4\frac{1}{2}$ metres (13 to $14\frac{3}{4}$ ft.) high, and are mostly of good size, excepting those in the topmost storey. Very few have chimneys. The walls are papered, partly with washable, partly with ordinary paper. The floors are everywhere of polished wood, or else covered with linoleum. Carpets are only used in strips by the bedside. Curtains and hangings are all of washable material. The furniture is mostly simple in character, but varies according to the kind of room. Stuffed articles have washable covers to them. There are no special ventilating contrivances, the air being renewed by open windows. Nor is there any lift. The place is heated partly by low-pressure steam, partly by ordinary German stoves or hot-water pipes. There are altogether six heating furnaces. Until recently the larger rooms were lighted with oil-gas, made in a private apparatus close to the sanatorium; the bedrooms, with lamps and candles. Electric lighting has, however, now been introduced, thereby effecting a much-needed improvement. The water supply, which is abundant and of excellent quality, is obtained from private springs at the top of the neighbouring hill. In case of drought, there is a second supply from Cronberg. Hot and cold water are laid on to each floor; but the supply of bathrooms was far too limited when I visited the institution. The drainage system, which was constructed in 1883, under the direction of an English engineer, appears to be very good and complete of its kind. The water-closets, although not built out, are of good pattern. The sewage, with the rest of the waste water, runs into precipitating tanks in a shed in the grounds, where it is treated with acidulated aluminium sulphate, and repeatedly clarified by a partially automatic apparatus. The sludge is mixed with peat, leaves, or earth, and then used in the garden, while the liquid effluent irrigates a field. There was no objectionable smell when I visited the place. Most of the outbuildings—the stables, cowsheds and pigsties, the somewhat primitive laundry, the steam disinfector, etc.—are on the other side of the road, behind the building on the north side.

There is altogether accommodation for 112 patients in the sanatorium, without counting those who lodge in the village. The proximity of the village inn, although in some respects convenient, is in others a disadvantage, or, possibly, in certain cases, a hygienic danger.

The treatment of the patients is according to Dr. Dettweiler's well-known methods. More stress is laid upon rest in the open air than upon graduated exercise, for which, indeed, the grounds are less suitable than at some other establishments. There is plenty of provision for open-air treatment, the deep verandahs, balconies, summer houses, and long covered corridors being available in all weathers. One of the corridors, indeed, is glazed, and can be heated to any desired degree, while an abundant supply of fresh air can be admitted. Hydrotherapy is systematically made use of, beginning with dry rubbing and passing through spirit rubbing, cold ablutions, and the cold wet sheet to the cold douche. The number of those who use the latter is, however, much smaller than formerly. Patients are somewhat unnecessarily restricted in their use of hot baths, being only allowed to have them at intervals, and with water below 96° F. Food is abundant and of good quality, four *meals* per day being provided ; the patients can also obtain extra milk by paying for it. An attempt, which is only partially successful, is made to suit the different requirements of the various nationalities represented amongst the patients. The milk is obtained from a private herd of tested cows under veterinary supervision, and is strained through felt. Dr. Dettweiler's known predilection for the administration of comparatively large doses of strong spirits is contrary to prevailing medical notions.

There is a strict rule enforcing the use of Dettweiler's flasks, or of ordinary *spittoons*, many of which are placed in the grounds. In the latter place they seem to me calculated to lead to just those evils which they are intended to prevent. A solution of lysol is used for disinfection. The linen of the establishment is washed by hand in the private laundry, after disinfection in the steam disinfector. Patients' body linen and handkerchiefs are washed in the village. Rooms are carefully disinfected after a patient leaves. There are no trained nurses, unless they are specially ordered from Frankfort, in which case there is no convenient accommodation for them. The servants' quarters are also, as I am informed, not quite satisfactory according to English notions, and apt to lead to awkward consequences in the event of accidental illness. Drugs are only sparingly used at Falkenstein. Creosote is little used; Koch's new tuberculin was being sparingly used when I was there. Great attention, however, is paid to throat complications, which are treated by inhalations and various topical applications. Various local societies amongst the patients help to keep them amused; but no regularly recurring entertainments from outside have hitherto been arranged. There is a Roman Catholic church in the village. Protestant services are held twice a month in the large board room. There is also a synagogue in Falkenstein.

The charges vary from 63 to 108 marks per week; reductions for children and servants. The entrance fee is 20 marks. Baths, douches, rubbing, drugs and instruments, beverages, afternoon milk, lighting, and, in the main block, heating, are extra charges.

The sanatorium is exclusively reserved for consumptives.

About 160 patients are admitted for treatment. The average for the last few years has amounted to from 400 to 450 patients per annum. The medical officers, who are well known for their courtesy and ability, are Dr. Hess, Dr. Besold, and Dr. Adolf Koch.[1] Dr. Dettweiler, the consulting physician, lives in Cronberg.

In the foregoing description I have been compelled to call attention to a few small blemishes in this historic and most valuable institution. That they have not greatly diminished its utility is evident from the statistics published of patients treated there (see p. 52), and from the high reputation which it enjoys. I have mentioned these defects both in the interests of accuracy and in the hope of strengthening the hands of the medical officers, who are always endeavouring to improve the institution, but have been hampered by exceptional difficulties in introducing some of the recent improvements. Falkenstein was for years the only sanatorium in western Germany for the rational treatment of consumptives, and has had a great influence in promoting the erection of other sanatoria, both in Germany and in other countries. For this reason, the names of Dettweiler and of Falkenstein will always be held in honour by practical physicians all over the world. To Dettweiler also belongs the credit of having brought about the erection of the neighbouring sanatorium for the poor, which was the first of its kind in Germany (see p. 253).

Falkenstein may be reached by carriage from Cronberg, which is the terminus of a short railway line from Frankfort, and about forty minutes' journey from that city. The sanatorium can also be reached by carriage from Soden railway station.

THE HOHENHONNEF SANATORIUM

belongs to a company whose shares are held by the inhabitants of a large part of western Germany, and of even more distant parts. It was opened in October, 1892, under the

[1] Since resigned ; see p. 196.

UPPER FLOORS

GROUND FLOOR

BASEMENT

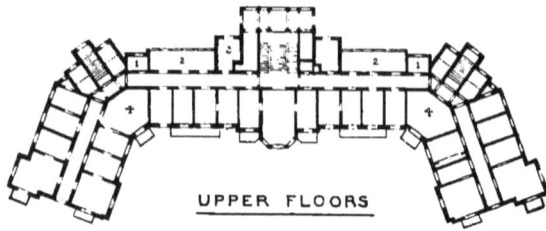

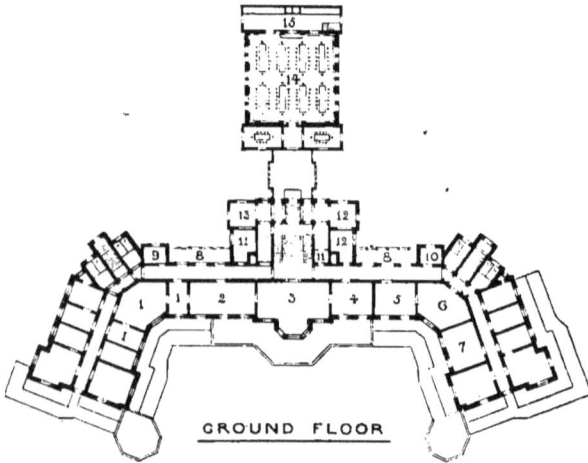

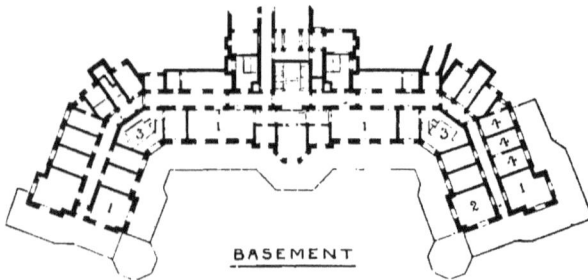

FIG. 21. THE HOHENHONNEF SANATORIUM. [*Face page* 183.

Basement :—

  1. Cloakrooms.
  2. Inhalation Room.
  3. Heating Apparatus.
  4. Douche Rooms.

Ground Floor :—

1. Consulting and Waiting Rooms.	8. Cloakrooms.
2. Winter Garden.	9. Servants' Room.
3. Reception Room.	10. Nurses' Room.
4. Reading Room.	11. Hairdressing Room.
5. Ladies' Room.	12. Office.
6. Music Room.	13. Post Office.
7. Billiard Room.	14. Dining Saloon.

     15. Serving Room.

Upper Floors :—

  1. Nurses' Room.
  2. Balcony.
  3. Small Kitchen.
  4. Rooms for Convalescents.

Fig. 22.—View of Hohenhonnef Sanatorium.

[*Face page* 183.

auspices of Dr. Meissen, who was formerly one of Dr. Dettweiler's assistants, and is situated in the Siebengebirge, 236 metres (774 ft.) above the sea-level, and 158 metres (518 ft.) above the Rhine, in a beautifully wooded and picturesque district, with fine views of the distant Eiffel mountains and the Rhine valley. The climate of this district is mild and sunny, but is not so dry as that of central Germany. The soil is Lower Devonian, composed of hard rocks at a considerable angle, so that it remains very dry. The grounds cover about 40 hectares (98¾ acres). of land, consisting partly of old and newly planted pine woods, partly of garden, partly of vineyards, and are continuous with a large extent of similar land on three sides.

The sanatorium, which is built of white stone and plaster, makes a very conspicuous landmark above the river side. It has a south-westerly aspect, and is well sheltered by higher hills (from 300 to 700 ft. higher) on the north and east, but is still somewhat exposed to south-westerly gales. However, these are not frequent, while the pine trees will in a few years afford better shelter.

The sanatorium consists of a five-storey building (figs. 21, 22), with two wings which diverge at a somewhat more open angle than those at Falkenstein. Owing to the slope of the hill, the basement is on the ground level in front, while the entrance behind is on a higher floor. The basement is occupied by cloak rooms, bath and douche and inhalation rooms, furnaces for heating the building, and a few bedrooms for visitors. On the same level, in front and round the eastern wing, is a deep verandah with cemented floor, which can be protected when necessary with glass, and part of which can be cooled in summer by streams of water on the roof. This is used as a fresh-air gallery, and is provided with cane lounge chairs, electric lights and bells, and small tables. Above it is an uncovered balcony. On the ground floor of the building are a series of intercommunicating handsome reception rooms, consulting rooms, laboratories, and drug rooms, as well as a few bedrooms for patients, and near the entrance the bureau and post office. The upper floors contain patients' rooms in a single row along the centre, a double row in the wings. There are large balconies both front and back, the northerly ones being specially frequented in summer during the heat of the day. Bath rooms, water-closets, rooms for attendants and nurses, and a large room for convalescents, are found on all the upper floors, which are served by a lift for patients, as well as staircases. The patients' rooms are nearly all on the sunny side of the building, and are well ventilated. Each is 4 metres (13 ft.) high, and has large windows and a chimney ; while none have less than 70 cubic metres (nearly 2500 cub. ft.) air space. The windows are double, for extra protection in stormy weather; the doors are also double, to prevent noise. Great care has been taken in constructing and arranging the sanatorium so as to facilitate thorough disinfection and cleansing without raising dust. Angles and corners are rounded, projecting cornices and mouldings avoided. The floors are made of "gypsdielen," consisting of bamboo and other light materials, incorporated with plaster of Paris, and are covered with linoleum or parquet. The walls are papered, as it is contended that this facilitates natural ventilation through their pores. Carpets, curtains, and the like, are sparingly used, and chosen of kinds which do not hold the dust and are readily cleaned. The furniture also is simply constructed but comfortable ; leather-covered chairs, cane lounges, and the like. The whole building is heated with low-pressure hot-water pipes, which can be separately regulated in every room, and is lighted by electricity. Even the summer houses in the woods and the fresh-air galleries are provided with electric light and electric bells. The dining saloon—a large, handsome, well-lighted room—is situated in a detached block to the north, reached by a bridge from the first floor, and has no direct communication with the kitchen department and storage rooms beneath. The servants sleep in a separate building in the wood behind the sanatorium.   The water supply, which is excellent, comes from a deep well in the Asbach valley, about half a mile behind the institution, where also are found the steam laundry and disinfector, and the engines for pumping and electric lighting.   The sewage and waste water are carried into the Rhine after precipitation in settling tanks with turf and coke bottoms, the solid parts being mixed with peat and dried in sheds, and afterwards applied to the land.   At high tide the waste water is made to irrigate a meadow.   Kitchen refuse is buried with ashes. *Sputa* are received into Dettweiler's flasks and elegant coloured spittoons, which are disinfected with · solution of 5 per cent. soft soap and 1 per cent. lysol.   The linen is disinfected by steam.   After each patient's departure, the walls are cleaned with bread crumb, which is then burnt.

In treatment, Dettweiler's methods are mainly adopted, with certain modifications—patients taking rest or exercise according to circumstances.   There are now two fresh-air shelters in the woods, and a third is to be constructed. Some of the walks in the pine wood are very well sheltered, and the fresh-air galleries next the sanatorium can be used for both rest and exercise.   One portion, indeed, can be warmed in winter or cooled in summer.   There is little or no douching of patients, but rubbing with the wet sheet or with spirit is commonly adopted.   The food is exceptionally good, five *meals* a day being provided.   The milk is supplied from a model dairy farm belonging to one of the directors.   Creosote, tuberculin, and other reputed specifics are not used.   There are three male and three female nurses, one of each on each floor.   A small Roman Catholic chapel is to be found in the grounds, where service is held once a fortnight.   For Protestants, service is held once a fortnight in the ladies' saloon.   There is a good library, and in other respects patients are well entertained and looked after.   Concerts are held about once a fortnight.

Separate houses for the chief medical officer and the managing director have been recently built close to the western wing. This has set free additional rooms in the sanatorium, so that there are now ninety beds, and very shortly there will be 109 beds in eighty-eight rooms, besides six private sitting rooms. During 1897, 798 patients were received, with 31,325 days of treatment. The charges are from 66 to 91 marks per week, with a reduction in case of friends, servants, and children. The entrance fee is 15 marks, and another 6 marks are charged for disinfection and insurance. Beverages, baths, rubbing, douches, inhalations, laundry, table lamps in the bedroom, attendance at night, and meals in the bedrooms, are extra.

The medical director is Dr. E. Meissen, formerly Dr. Dett- weiler's assistant. There are two assistant medical officers. —Dr. Schröder and Dr. van Ysendyck. The *personnel* now number seventy-nine.

This is probably the most luxurious sanatorium for con- sumptives on the continent. There is, indeed, a little danger lest the internal comfort should tempt the patients. to spend too much time indoors; but I saw no indications. when I was there of such mistaken conduct, which is no doubt prevented by Dr. Meissen's watchful care. There are fewer facilities for uphill walks from the sanatorium than might be wished, and the shelter from certain winds is not yet sufficient; but in almost every respect it is a model establishment, marvellously complete in its details.

It may be reached by the right Rhine railway, taking carriage from Königswinter or Honnef; or by the left Rhine railway to Rolandseck, and thence by ferry to Honnef; or by steamer to Königswinter or Rolandseck. The institution is twenty minutes by road from Honnef, or forty from Königswinter.

THE LAUBBACH SANATORIUM.

About twenty-five minutes' walk southward from Coblenz, on the left bank of the Rhine, in a sheltered cleft surrounded on three sides by mountains, is situated an in- stitution which was originally a hydropathic establishment, but has been recently acquired by Dr. Achtermann (formerly medical director of the Brehmer sanatorium), and converted into a sanatorium for consumptives. About $8\frac{1}{2}$ hectares (21 acres) of land belong to the establishment, which are mostly wooded with deciduous trees, and touch the town woods of Coblenz. The well-known castle of Stolzenfels is a short distance off, and the pleasure gardens by the side of the Rhine are also close by. The soil is rocky, and in the sanatorium grounds is free from dust. Owing to its sheltered position the place is said to be 2° to 3° R. cooler in summer, and about as much warmer in winter, than the immediate neighbourhood. The climate resembles that of the Rhine valley generally, and is mild and moderately humid. The shelter against wind is extremely complete.

The sanatorium is 80 metres (262 ft.) above the sea-level, and some 20 or 30 above the Rhine. It has been a health resort since 1840, but was rebuilt in the eighties by Dr. Auerbeck. Passing up the avenue of chestnut trees which form the approach, we come to an open square with a few smaller trees in the centre and several large blocks of buildings to right and left. On the right is the bath house, with various hydropathic appliances, the office and consulting room, a winter garden, and two more large day rooms.

On the upper floor are thirteen patients' bedrooms. On the opposite side of the square is the main building, with fifty-four bedrooms on the upper floors, and a handsome lofty dining saloon on the ground floor, and near this a light and airy kitchen. The windows of the dining saloon are large; in front they look on to the open courtyard; at the back they are semi-opaque, to shut out the sight of the cliff. Touching the main building is the " doctor's house," which has fourteen patients' bedrooms, and in the basement a smaller dining saloon for the "first-class patients". Next this is a villa with seventeen bedrooms, and a little higher up in the grounds are two others with fifteen rooms, thus making a

total of 113 bedrooms available for patients. The rooms
vary in size, but are all lofty, and average 80 cubic metres
(2825 cub. ft.) air space. They are furnished and decorated
in much the same style as the average hotel of the Rhine
district, and are all single bedded. The buildings are heated
by hot-water pipes and by closed stoves, and lighted by the
Coblenz gas supply. The ventilation is by open windows;
there is usually no chimney. The sewage is carried through
the Laubbach into the Rhine. There is a chapel on the
grounds; also several shelters for the fresh-air treatment,
besides protected seats under the trees. The patients are
well provided with means of amusement, as there are a
billiard room and a bowling green, in addition to the
library of books, etc.

The establishment is not entirely reserved for consumptives, as those in need of change of air are also admitted.
No hopeless or advanced cases are accepted, but there are
two sick nurses in the establishment in case of need.

Treatment is on the usual lines. Laryngeal, pharyngeal,
and nasal complications are said to do better than in more
elevated districts. The usual five *meals* a day are provided.
The douche is freely used in all cases which are sufficiently
robust. Wine and spirits are prescribed when advisable.
No cod-liver oil or reputed specifics are employed. *Sputa*
are received into Dettweiler's flasks; they are kept moist
until they can be poured into the water-closet, but are not
disinfected. Linen is boiled; walls purified by bread
crumb, which is afterwards burnt.

The charges are 6 marks per day, including attendance,
rubbings, douches and baths, and ordinary medical attendance. Rooms are 1·50 to 5 marks per diem. Another 2
marks per week are charged for the use of bedding, including the washing of bed linen; but patients may, if they
prefer, bring their own. There is an entrance fee of
20 marks. Beverages are extra, and in winter 5 marks
per week are charged for heating and lighting. Reductions
are made for visitors, children, and attendants.

SECOND-CLASS PATIENTS

are also received into the sanatorium. Their food is somewhat simpler, but the treatment is otherwise the same.
They pay an entrance fee of 12 marks, and 145 to 160
marks per month for board and lodging, including everything excepting beverages, medicines, and bedding. For
the latter, 1·50 marks per week are charged. In winter,
an addition is made of 3 marks for heating and lighting.
There are reductions for visitors and attendants, as in the
case of "first-class patients".

This sanatorium is at present in a state of transition, and
will in (perhaps) the near future be considerably improved.
Dr. Achtermann has been put to considerable expense in
making the absolutely necessary alterations; but with his
long experience at Görbersdorf to help him, he is sure to
eventually bring the establishment into the first rank.
The first- and second-class patients are lodged in separate
buildings, but are close together. As already explained,
they have separate dining saloons.

The nearest railway station is the Moselbahnhof at
Coblenz.

CHAPTER XXXI.

THE sanatoria in this district for paying patients are the Nordrach Colonie, Dr. Sander's sanatorium at St. Blasien, Dr. Koch's at Schömberg, and Dr. Leiser's at Badenweiler. Other sanatoria in the same district for poorer patients are described in chap. xl.

THE NORDRACH COLONIE

is about 30 miles from Strassburg, and 4½ miles from the little village of Nordrach, in the Grand Duchy of Baden. Leaving the little station of Biberach, on the railway from Offenburg to Constance, and following the road through Zell and up the valley of the Nordrach, which is dotted with picturesque wooden peasants' houses, a small cluster of cottages is met with at the bifurcation of the valley, which marks the position of the disused glass factory now replaced by Dr. Walther's sanatorium. The place is sheltered on all sides excepting the south-west, by pine-clad hills, which gradually slope to a height of about 900 metres (3000 ft.) on the north and east, the sanatorium itself standing at a height of about 450 metres (1470 ft.) above the sea-level. This spot was chosen by Dr. Walther for his sanatorium, after many months' searching, as combining all the important features necessary to such a sanatorium. The subsoil is a coarse red sandstone, which rapidly absorbs moisture ; the wind shelter is very complete ; a southerly aspect with an open valley and lower hills to the south ensure plenty of sunshine, while the distance from the nearest town (15 miles) and the nearest railway station (9 miles) ensures rest and tranquillity for the patients. The air is pure, and the river Nordrach ensures an abundant supply of good water and a never-failing source of power. The climate is that of a moderately elevated hill district—somewhat variable and moist as compared with Alpine climates, but with a fresh, still air, and a large proportion of rainy days. According to Dr. Rowland Thurnam,[1] it is somewhat colder in winter and hotter in summer than that of the English south coast, and is visited at times by heavy rain-storms, while in winter there are severe frosts and snow, interspersed with occasional gales of wind, and a certain amount of mist and dull weather. It is very rare to see any dust blown about—an obvious advantage to phthisical patients. About eighty acres of ground belong to the sanatorium, which is further surrounded by miles of forest land, traversed by a network of paths. The grounds are provided with paths at every variety of gradient, with seats at intervals, but with only three small huts or shelters, and no regular "liegehallen" or covered walks. This arrangement is made of set purpose, as Dr. Walther believes that when rest is needed it is best taken in isolation, in the patient's own bedroom.

The sanatorium was originally started in 1888, when it consisted of a single house with accommodation for ten patients. It now has rooms for forty-five, besides a small villa which is being built for old patients who may come on a visit.

The sanatorium consists of detached buildings, none of which have more than two floors. The office, kitchen department, and rooms for the staff, are situated in the old *Gast-haus* (or inn) *zum Anker*, which is connected by telephone with the railway station at Biberach, and with the patients' houses. Attached to this building is the dining saloon, a light and cheerful building about 40 by 20 feet,

[1] Harris and Beale, *The Treatment of Pulmonary Consumption*, London, 1895, p. 376.

with large windows back and front, linoleum-covered floor, and varnished panelled wooden walls. These windows are taken out of their frames during the warmer part of the year; those in front look out upon the road, those behind on the side of the hill. The dining saloon and a small ante-room with piano and library constitute all the common rooms in the sanatorium, so that there is every inducement for the patients to spend their time out of doors. The houses are not connected in any way, so that the road has to be used in going to and fro. The main building is situated on the slope of the opposite hill, some 20 or 30 feet higher than the dining saloon. It is a plain wooden building with a little turret at each end, consisting of a single row of bedrooms on the ground floor, with corridor at the back, and a similar set of rooms above. The main aspect is south-east, and the ends of the building retreat slightly to follow the convex face of the hill. The patients' bedrooms are not particularly large ($3 \cdot 5$ metres $\times$ $4 \cdot 5$, and $2 \cdot 7$ high, or 11 ft. 6 in. $\times$ 14 ft. 9 in., and 8 ft. 10 high); but they have a relatively large window space amounting to about half the size of each room, and the windows and doors are permanently kept open. The walls and ceilings are entirely lined with plain varnished wood, which is revarnished whenever a patient leaves. The floors are covered with linoleum, which is washed down every morning. In one corner of each room is a shower bath or douche apparatus; in another is a plain iron bedstead with woollen bedclothes; and the remaining furniture is of lacquered wood or iron, and of equally simple character. There are no carpets or hangings. Hot and cold water are laid on for the douche and the washstand. The corridor is similarly built with pinewood panelling, which here is oiled. The windows of the corridor are opposite the bedroom doors. The water-closets, which are built out of the corridor, are provided with a good flush of water. Twenty patients can be accommodated in this building. The rest are lodged in other villas farther up the valley. Dr. Walther occupies one of these.

Another villa is occupied by eleven slight cases; this is the only existing building in the sanatorium which has a common bath-room for the patients. In all essential respects, the other villas are constructed and furnished like the main building. There are no special ventilating contrivances. The lighting throughout the sanatorium is by electricity generated by a motor driven by water power; all the rooms, too, have electric bells; and one entire villa is heated by electricity. Elsewhere the sanatorium is heated by low-pressure steam pipes, which can be regulated for each building as a whole. Passages and closets are not separately warmed. There is a good and plentiful supply of drinking water from springs out of the sandstone rock. The sanatorium possesses its own farm of ten cows, which are fed on barley and hay only. The milk for cooking is, however, obtained elsewhere. I believe, though I unfortunately forgot to ask Dr. Walther about it, that he considers it unnecessary to test the cows with tuberculin. Patients have the milk fresh or boiled, whichever they prefer. There is a steam laundry in the sanatorium, and a mechanical refrigerator for meat and for ice. All meat is kept a fortnight before being used.

Some of the descriptions of Nordrach Colonie give rather a wrong impression of the place, as if it merely consisted of ordinary peasants' houses and were devoid of all comfort. No attempt is made, it is true, to provide more than ordinary comfort, or to rival the luxury and sumptuousness of a first-class modern hotel. But these are not at all necessary, and in some respects are misplaced in a sanatorium. The rooms, as far as I saw them, were exceedingly carefully and practically, though simply, fitted up, and were exquisitely clean and sweet, which is more than can be said of one or two other sanatoria that I have seen. Opinions may differ as to particular details of Dr. Walther's system of management, but for all that he does or omits he has excellent reasons to give.

Open-air treatment at all hours, weathers and seasons is
13

very rigidly enforced, extra wraps being provided in bad weather, but no other difference being made. Patients walk about in even rainy weather without any special protection. They take their own temperatures four times a day *in recto* as a guide to treatment. Dr. Walther regards all other methods as fallacious and misleading. Until for eight successive days the temperature has failed to reach 38° C. (100·4° F.) patients are kept absolutely at rest in bed with open windows. They are then permitted to take very gradually increasing exercise, mainly walking, never faster than two miles an hour, so as not to cause dyspnœa or disturb the circulation. The first walk may only consist of a few yards on level ground, followed by a long rest on a lounge chair in the open; afterwards shorter or longer uphill walks are prescribed, according to the progress made. Patients are encouraged to do without heavy overcoats and wraps while out walking, although they may wrap up as much as they like while resting. Not more than three or four are allowed to walk together, and visits from friends or relatives are discouraged. Even attendance at the little chapel at Nordrach is objected to, although there is no regular service in the institution. Dr. Walther believes weather and season to be of no import- ance, and does not even take meteorological observations. Nor are any special precautions adopted against draughts. As he himself told me, patients live in a thorough draught from the first moment of their arrival, with absolutely no bad effects, although no attempts are made to " harden them " against catching cold. Hydrotherapy is not syste- matically employed with this object, but patients bathe or douche themselves with hot or cold water, very much as their own instinct leads them, and in clothing are also allowed to please themselves within certain limits. They rest every day for an hour before the mid-day and evening meals.

Several inaccurate accounts have been given concerning the *diet* at Nordrach, even by presumably trustworthy ob- servers, and Dr. Walther was at some pains to correct these inaccuracies. Three meals a day are provided, at 8·30 A.M., at 1 and at 7 P.M. The breakfast consists of coffee, bread and butter, and cold meat of some kind. Dinner includes two hot courses of meat, or fish and meat, with plenty of potatoes and green vegetables, and sauces containing butter. Following this are pastry, farinaceous pudding, fruit or ice cream, with coffee to finish. Supper consists of one hot and one cold course, together with tea. Milk is added to the dietary until the patient's weight has reached a reasonable standard, never more than a half-litre being given with each meal. The meals are taken in the presence of the doctor, who encourages the patient to finish what is given him, and eat a reasonable quantity of every kind of food provided. The stories about patients being compelled after vomiting to begin their dinners over again against their will are inaccurate. So also are the statements as to enormous quantities of meat or food generally being given. The proportion of nitrogenous food given is if anything less than usual; and no attempt is made to emulate Debove and others of the same school in hyper-alimentation. The quantity is regulated so as to produce a steady gain in weight during the earlier weeks of about one lb. per week. Under this regime dyspepsia is said not to be common, and soon to disappear, if present, with improving nutrition and strength. The use of alcohol is kept down to a minimum, without any absolute prohibition.

Drugs are but little used, cod-liver oil, creosote and reputed specifics not being employed. *Sputa* are disinfected with liquor potassæ, and then poured down the water-closet. Each patient has his own Dettweiler's flask for the pocket and a spitcup in his bedroom, and spitting elsewhere is strictly forbidden. Linen is disinfected by boiling; the rooms by washing and airing and revarnishing. There are practically no amusements, excepting walking, reading, and occasionally a little music. Light literature is permitted, but business communications or work of any kind are forbidden.

Monthly examinations are made of the lungs and expectoration. When after about twenty examinations no bacilli
are found in the sputa, and when after injection of one c.c.
of the sputum into a guinea pig tuberculosis does not appear
within three to six weeks, the patient is considered fit to
return home, and to undertake moderate regular work.

Dr. C. Reinhardt writes:[1] "All nationalities seem to be
represented, the whole company presenting a lively, well-
dressed, and happy appearance. I spoke with many
patients . . . and did not find one who had not gained in
weight or who had suffered from dyspepsia or nasal catarrh
after the first month. Every case had experienced improvement, and several were apparently cured. The friend
of mine, a medical man, whose cure had first directed my
attention to the place, remains well and is in active pursuit
of his profession, although, before his visit to Nordrach, he
had been in an advanced phthisical condition, with a large
cavity in the right apex and much general emaciation,
whilst his sputa had teemed with bacilli."

The charges, which include everything excepting washing, alcoholic drinks, and the services of a special nurse,
amount to 10 marks per diem; friends pay 7 marks.
There are arrangements for the schooling of children during their stay. Dr. Walther is aided in his work by an
assistant.

The place is nine miles from Biberach-Zell, and ten miles
from Gengenbach, on the Black Forest Railway, whence it
may be reached by carriage.

THE SCHÖMBERG SANATORIUM

is situated in Würtemberg, in the northern part of the Black
Forest, south-east of Pforzheim, and not very far from
Wildbad. A convalescent home was erected here in 1888
by Herr Römpler, which, being much frequented by consumptives, was converted in 1890 into a sanatorium exclusively for such patients, and placed under the care of

[1] *British Medical Journal*, 7th August, 1897.

Dr. Baudach. It is now under the medical direction of Dr. Adolf Koch, formerly of Falkenstein, Dr. Baudach having resigned. There were originally only twenty-five beds ; but in 1893 a large three-storey building was erected, with forty beds, when the older building was utilised for the reception of less wealthy patients (see p. 260).

The village of Schömberg lies in a valley on the mountain crest, between the Enzthal and the Nagoldthal, at an altitude of 650 metres (2130 feet) above the sea-level, and surrounded by pine-clad heights to the north, east and west. The sanatorium is on the southern slope of a hill in the midst of the village, but in a fairly open situation. It has about 1½ hectares (3½ acres) of ground, partly laid out as a garden. The paths all lead uphill towards the woods, which are distant about ten minutes' walk. The soil is of porous sandstone, as in other parts of the Black Forest ; the climate in summer cool and pleasant, without excessive evening fall of temperature, and in winter is for the most part sunny, the snow remaining the whole season on the ground.

The building (fig. 23) has a south-south-east aspect, all the patients' rooms being on the south side. It is of oblong form, with a deep verandah along the front, and balconies to every other floor but the uppermost. In the basement there are a douche room, a large steam disinfector, laundry, kitchen and central heating apparatus for low-pressure steam pipes. The ground floor contains the dining hall and other assembly rooms, such as the reading room, billiard room, and ladies' room. The largest of these contain respectively 250 and 320 cubic metres (8825 and 11,300 cub. ft.) air space. On the upper floors are the bedrooms, and on each floor a bath room and servant's room. The rooms have an average capacity of 60 to 80 cubic metres (2120 to 2820 cub. ft.) ; their walls are oil-painted for 1¼ metres from the ground near the beds, colour-washed elsewhere. They are ventilated by air shafts, and by open upper hinged window panes, and decorated simply but comfortably. The heating is by low-pressure steam ; the lighting partly by benzine incandes-

cent lights or petroleum, partly by means of candles. There
is a good water supply. The water-closets lead into
settling tanks.

The treatment is much as at Falkenstein, with a large
proportion of rest in the open air. Besides the fresh-air
verandah, which is intended mainly for those whose rooms
have no balcony, there are several shelters in the grounds,
in one of which, on the edge of the wood, patients can
obtain milk, and enjoy a beautiful view of the surrounding
country. There are five *meals* per day, in addition to
milk before bed time. The douche and cold rubbing are
frequently used ; cod-liver oil if desired ; "specifics" are
not administered. *Sputa* are received into water but not
disinfected ; linen is purified by current steam ; rooms with
creolin and formalin. Only one patient sleeps in each
room. There are one male and one female nurse ; besides
which nursing sisters are obtainable in case of need. The
general management is in the hands of Herr Römpler,
himself an example of recovery from pulmonary disease.
The medical director, Dr. Koch, who lives in another house
close by, is assisted by Dr. Wehmer, formerly at Görbersdorf. The latter lives in the sanatorium.

A good library, piano, croquet and other games are provided for the patients. There is a Protestant service in the
village church ; no Roman Catholic service is available.
The charges are from 32½ to 57 marks per week, in addition
to 5 marks for the first consultation, 1·50 for attendance,
1 mark for the bath attendant, and extra for beverages,
douches, baths and drugs. The usual reductions are made
for friends, servants and children.

Schömberg is a telegraph station, and is on the telephone.
It may be reached by carriage in 1½ hours from Liebenzell
(on the line from Pforzheim to Calw), or from Höfen (on
the Wildbad line). Heavy luggage comes *via* Höfen.
Pforzheim is not far from Carlsruhe.

There is accommodation in the old and new Curhaus
together for 140 patients (see p. 260). The average number

of patients treated during the four years 1893-96 was 358, with 10,324 days of treatment (Hohe, *Die Bekämpfung und Heilung der Lungenschwindsucht,* Munich, 1897).

DR. SANDER'S SANATORIUM,

in St. Blasien (Baden), was founded by Dr. Haufe, in 1878, but has been since 1895 under the management of Dr. Albert Sander.

St. Blasien is an ancient and picturesque little town in the valley of the Alb, in the southern portion of the Black Forest, famous for its eleventh century cloisters and church, and boasting of a flourishing kursaal and hydropathic establishment. The district is on primary rocks, and possesses an equable climate, with fairly cool summers and mild winters. The town itself, however, is somewhat shut in, as, notwithstanding its elevation above the sea-level (800 metres, or over 2600 ft.), it is surrounded on all sides by mountains, one of which—the Feldberg—is the highest peak in the district, and rises to a height of 4900 feet. This circumstance renders the place unpleasantly hot in the height of summer, although in winter it is probably much more agreeable. The drainage is also somewhat primitive, and spoils the charm of what would otherwise be at certain seasons a most desirable health resort.

Dr. Sander's sanatorium is at the western end of the town, on the pine-clad valley side, a little raised above the main road, and nearly 100 feet above the lowest part of the town. It has a southerly aspect, and consists of three separate buildings, united by a glass verandah 70 metres long, and provided with terraces, balconies, and other verandahs for rest in the open air. Behind it rises the forest, which is traversed with paths and has a large shelter among the pine trees, for use in bad weather. The old building has kitchen and disinfecting apparatus in the basement, two dining rooms, several assembly rooms, waiting and consulting rooms and the office on the ground floor; and, above this, two floors with twenty-six bedrooms for patients. The new villa has fourteen patients' rooms, as well as those for Dr. and Mrs. Sander and family. Additional accommodation can be obtained in a third building touching the sanatorium, but belonging to a different owner. The rooms are four metres (13 ft.) high, and are comfortably furnished, with papered walls, parquet floors, and linoleum in the corridors. Carpets are only used as strips by the bedside. The sofas and stuffed chairs have washable covers, and the curtains are also washable. The common rooms have air shafts for ventilation ; the bedroom windows have ventilating panes above, which can be separately opened. There are bath-rooms and covered verandahs on every floor, and lifts for wood and food. The heating is mainly by means of tiled stoves, in which beechwood is exclusively burnt. The corridors and staircases have regulating anthracite stoves, the closets being also warmed. The latter are mostly of a new pattern, but are connected with cesspools. However St. Blasien is now being sewered, so that the sanatorium will in due course profit by the sanitary improvement. The whole establishment, including the "liegehalle," is electrically lighted. The well-kept garden is directly continuous with the pine wood behind the sanatorium. There are two bridges from the second floor of the building, leading directly into the woods.

Treatment is by the usual methods. Patients are kept indoors when the thermometer falls below − 7° C. *Sputa* are collected in elegant spittoons and disinfected by three per cent. solution of lysol. Patients' linen is washed in the town. There are five *meals* a day. There is plenty of amusement for the patients, both in the establishment and (during the season) in the town. Indeed Dr. Léon Petit rightly regards the proximity of the kursaal as a serious drawback to the treatment. There is at present no douche room in the sanatorium, so that patients have at times been sent to that in the town. But this defect will probably soon be remedied. There are Protestant and Roman Catholic churches in St. Blasien.

The charges are from 68½ to 82½ marks per week. Wood for the bedroom stoves is extra, as well as beverages and an entrance fee of 10 marks.

There is accommodation for over sixty patients, fifty-one in the old and new buildings, and eleven in the *dépendance*. The average number of patients treated during the five years 1892-96 was 133, with 11,802 days of treatment. Patients are admitted at all stages, and stay from four to fourteen months. An assistant physician helps Dr. Sander in his duties.

The place is not very convenient of access, as it is distant 26 kilom. (16 miles) from Albbruck on the line from Basel to Constance, and 30 kilom. (over 18 miles) from Titisee on the Höllenthal Railway (Freiburg to Neustadt). This necessitates a drive of about four hours by omnibus or carriage.

Many of the foregoing particulars have been obtained from Hohe's book,[1] as Dr. Sander was unfortunately absent at the time of my visit, and I was unable to wait for his return.

DR. LEISER'S SANATORIUM.

It is stated by Möller,[2] Léon Petit,[3] and Kuthy,[4] that a sanatorium for consumptives exists at Badenweiler. Léon Petit refers to it as rather a boarding-house or convalescent home, where those in need of change of air, as well as early stage consumptives, are received. Both he and Möller state that it was to be rebuilt and replaced by a larger establishment; and Kuthy, who gives a fuller description, and states that it is under the care of Dr. Leiser, also mentions the projected rebuilding. I was unable to find such an establishment at Badenweiler itself, but was told of one being built five or six miles farther east, at Marzell, on the Blauen, which time did not permit me to visit.

[1] *Loc. cit.*

[2] *Les Sanatoria pour le traitement de la Phtisie Pulmonaire.*   Brussels, 1894.

[3] *Loc. cit.*   [4] *Loc. cit.*

Badenweiler itself is a small hamlet in the western part of the Black Forest, on the northern side of the Blauen, which rises to a height of 3830 feet. The village is 1385 feet above the sea-level, and enjoys a beautiful view across the valley of the Rhine towards the Vosges. It is well sheltered by mountains to the north, east, and south, and has a short and mild winter, the autumn lasting until December, and the spring beginning at the end of February. It possesses thermal waters which were known to the Romans, a kurpark and kurhaus, with concert, ball, and reading rooms, a handsome bathing establishment, and various Roman remains. A large number of visitors come every year for the whey cure and change of air.

Kuthy describes the sanatorium as lying in the midst of a fine park, wooded with pine trees. The dining room, three other reception rooms, and Dr. Leiser's own rooms, are on the ground floor. The patients' bedrooms are on two upper floors, served with a lift. Patients carry out the fresh-air treatment on balconies and in a pleasant pavilion in the garden. There is also a winter garden. On warm, sunny days, a terrace opening directly out of the dining room, with a northerly aspect, serves as a resting place. The heating is by stoves. Much milk is given to the patients, who are supplied with light wine, but not much spirits. As regards hydropathic treatment, tepid water is used at first, then colder water, then the cold wet sheet, and finally, in certain cases, the douche.

A sanatorium for the poorer classes is being built at Marzell, on the Blauen (see p. 262).

CHAPTER XXXII.

A NUMBER of societies have been founded in various parts
of Germany for the erection of popular sanatoria for consumptives. The earliest of these appears to have been the
" Verein für Gründung von Heilstätten für bedürftige
Lungenkranke," which was formed in November, 1888,
with committees in Hanover, Bremen, and Brunswick.
The first sanatorium, however, of this kind to be founded
in Germany was the one started in 1892 in a private house
close to the Falkenstein sanatorium by the " Frankfurter
Verein für Reconvalescenten Anstalten," mainly owing to
the efforts of Dr. Dettweiler (see p. 253). It was in the
same year that the Bremen " Heilstätten-Verein für
bedürftige Lungenkranke," which was an offshoot of the
earliest sanatorium society, began to build their sanatorium
at Rehburg (see p. 246). There are now in Berlin two chief
societies engaged in this work, the Berlin-Brandenburg
Society, with Prof. v. Leyden at its head, and the Red
Cross Society, under the patronage of Countess Hohenlohe
and the presidency of Dr. Pannwitz. The Berlin-Brandenburg Society, which was started in 1895, is intended to
promote the erection of sanatoria for the less wealthy
middle classes ; while the Red Cross Sanatorium Society,
an offshoot of the Red Cross Society (which is occupied with
philanthropic work of many kinds), caters for the lower
classes. A central committee was also formed in December,
1896, to co-ordinate the efforts of these and other similar

societies, with the Empress as its patron and the Imperial Chancellor as president. It publishes a sheet once a month (*Heilstätten Correspondenz*), edited by Dr. Pannwitz. A permanent committee for combating tuberculosis is also being appointed by the German Society of Naturalists and Surgeons.

Speaking in 1895 at the meeting of the German National Health Society, Prof. v. Leyden stated that there were in Germany only two sanatoria and two small houses where the poorer class of consumptives could be received for hygienic treatment. At the present time, however, there are nearly thirty already built, or approaching completion, while a number of others have been projected. This rapid progress has been mainly owing to the law enforcing compulsory insurance against sickness and old age on all whose annual income is under £150 per annum. This Act, which was passed in 1889, contains a clause permitting any sickness insurance society to devote part of its funds to the treatment of the sick in lieu of sick pay, and under this clause large sums of money have been expended in the erection and maintenance of sanatoria for the hygienic treatment of consumption. Within two years of the passing of the Act there were in Germany 11,200,000 persons insured against sickness and old age, and the various societies had collectively a capital of 162,850,000 marks. It has been estimated that half of the applications for sick pay between the ages of twenty and twenty-nine among the insuring classes in Germany are because of tuberculosis, so that it is not surprising to learn that in 1897 out of thirty-seven diferent sickness insurance and friendly societies, thirty-three had spent more or less of their funds on the hygienic treatment of consumptives. Altogether 4480 such patients were assisted by these societies in 1897, and of this number 4432 were treated in sanatoria or country colonies. In one way or another about 1,300,000 marks were invested in sanatoria during 1897 by these thirty-three societies; and for this year (1898) between three and four million marks have been set aside for the same purpose. In some cases money has been advanced on mortgage, in others the whole cost of erection has been borne by the society.

Gebhard, the director of the Hanseatic Sickness and Old Age Insurance Co., proved that if, out of 500 consumptives, 140 could be so far restored as to do without sick pay for a year, this would recoup the company for the cost of treatment in a sanatorium, or for the erection and maintenance of one of its own. This Hanseatic Insurance Co. is now one of several which possess sanatoria of their own. As a rule, the first step has been to send such consumptive patients to a suitable country village for treatment, or to pay for them at an existing sanatorium. After a while, when the numbers warranted such a step, the right to a number of beds has been purchased, or a private sanatorium erected by the company. Local societies for the erection of sanatoria have usually begun by spending part of the interest on money subscribed in maintaining a few patients at one of the various sanatoria, while the rest of the interest was added to the principal until such time as the funds were sufficient for the erection of a fresh sanatorium.

The popular sanatoria which have already been erected are for the most part described in the following chapters. A number of others, however, have been projected. Every large town, in fact, and every district of Germany now has its local sanatorium society, and often already its own sanatorium.

In the following tables the German popular sanatoria have been placed for the sake of convenience in two groups, one including eastern and central Germany, the other western and southern Germany. In the description which follows, a geographical order has been adopted which roughly corresponds with that in the chapters on private sanatoria.

GERMAN SANATORIA FOR THE PEOPLE.

EASTERN AND CENTRAL GERMANY.

		Feet.	Beds.
Görbersdorf, Krankenheim .	Silesia	1840	180
Loslau	,,	984	87 (100)
(Oberkaufungen for Cassel) .	,,	?	? 100
(Beelitz)	? Brandenburg	?	? 100
(Belzig)	,,	?	80-90
(Bleichröder, Belzig) . . .	,,	?	30
Blankenfelde	,,	?	60
Grabowsee	,,	125	80 (160)
Malchow	,,	102	86
(Stettin)	? Pomerania	?	80
Albertsberg	Saxony	2300	122
(Plauen)	,,	2130	?
Jonsdorf, Conval. Home .	,,	?	33
Oderberg	Harz	2100	115-120
St. Andreasberg, Felixstift .	,,	?	30
Stiege, Albrechtshaus .	,,	1380	58
(Stiege, Marienheim) . .	,,	1350	20
(Sülzhayn)	,,	? 1600	120
Altenbrak Colony . . .	,,	? 1300	?
Königsberg, Conval. Home .	,,	1470	? 36
Zellerfeld	,,	1800	40
(Berka Emskopf) . . .	Thuringia	?	abt. 80
(Manebach)	,,	1720	?
(Erfurt, Johanniter) . .	,,	?	?

NORTH-WEST AND SOUTH-WEST GERMANY.

		Feet.	Beds.
Rehburg (Bremen San.) . .	Hanover	? 320	30
Altena	Westphalia	1380	100
Honnef, Philomenen . .	Rhenish Prussia	? 260	?
(Dusseldorf)	,,	?	?
Ruppertshain	,,	1300	88
(Höchst am Rh.) . . .	,,	?	120
Dannenfels	Palatinate	1300	18
(Speyer, Alberswoiler) . .	,,	1250	?
(Worms, Felsberg) . . .	,,	?	?
(Saarbrucken, St. Arnual). .	Alsace-Lorraine	?	?
(Metz)	,,	?	?
(Würzburg, Lohr) . . .	Bavaria	?	30
(Nuremberg, Engelthal) . .	,,		60-100
(Planegg-Krailling) . . .	,,		114
(Harlaching)	,,		200 (350)
Stuttgart, Neustädtle . .	Wurtemberg	?	?
Schönberg, Colony . . .	,,	2130	60-65
Arlen	Baden	1380	16
(Marzell)	,,	2750	108

SEASIDE SANATORIA FOR CHILDREN.

The society for the erection of seaside sanatoria has been established for about eighteen years. It has four sanatoria : at Norderney (*Kaiserin Friedrich Hospiz*), at Wyk on the island Föhr, at Zoppot near Dantzig, and at Gross Müritz on the Mecklenburg coast. The Norderney Hospiz is the largest, and received as many as 149 visitors in 1894. These sanatoria are not specially for consumptives, and resemble our own convalescent homes.[1]

[1] *Brit. Med. Jour.*, 1st June, 1895.

CHAPTER XXXIII.

SANATORIA FOR THE PEOPLE IN EASTERN GERMANY.

ALTHOUGH Silesia was the original home of the Brehmer system of treatment, few sanatoria for the poor have been erected in Eastern Germany. A large open colony exists at Görbersdorf under Dr. Weicker, the proprietor of the Countess v. Pückler's sanatorium for paying patients (see p. 157). Another open health resort or colony for consumptives may be found at Wölfelsgrund, south of Glatz, near the Schneeberg; and this year a closed sanatorium for the poor has been opened at Loslau.

It is stated that in Silesia alone, out of 140,000 deaths per annum, about 11,000 are caused by consumption; so that the supply of sanatoria for the poor is evidently insufficient. A meeting was held this year at Breslau, with Count Hatzfeld in the chair, at which the erection of such a sanatorium for Silesia was decided upon; and a site is being sought in the Riesen and Eulen Gebirge. Quite recently a legacy of 30,000 marks has been left by a lady for the same purpose. A legacy has also been left for the erection of a sanatorium of 100 beds at Oberkaufungen for the *Vaterländische Frauenverein* at Cassel. Dr. Osius will be the medical officer.

In East Prussia a similar scheme has been originated by the local medical association; and at Dantzig, in West Prussia, money is also being collected for a sanatorium.

DR. WEICKER'S KRANKENHEIM,

at Görbersdorf, is not a " closed sanatorium," but consists
of a number of separate houses in the village, devoted
since 1894 to the treatment of consumptive patients of the
poorer classes. Only one of these—the so-called doctor's
house—belongs to Countess v. Pückler's bequest (see p.
157), and is occupied by an assistant physician under Dr.
Weicker, having been altered to suit modern requirements.
This house contains the kitchen, dining rooms, and administrative rooms, as well as beds for those who may be attacked
by intercurrent ailments. The other houses contain each a
common room, in addition to bedrooms for the patients.
Men and women are separately lodged, and have separate
collective dining rooms ; the women occupying eighty beds
in three houses. Nearly all of the villas have balconies ; but
there are also three large sun galleries for common use.
Many of the houses were built in Brehmer's time for a
better class of patients, but have nothing very unusual
in their mode of construction. Others are ordinary village
houses altered to suit their new functions. In every case
a garden surrounds the house. The rooms are described
by Hohe [1] as being lofty and cheerful. The dining-rooms
are on the south side. Walls are colour-washed or oil-
painted. Windows have washable blinds. The heating is
by means of porcelain closed stoves. Most of the bedrooms
contain each two to three beds, some being single bedded.
The beds are of iron, on wire mattresses, and provided with
over-mattresses. The bedding is of wool, changed every
week or fortnight. Every patient has his own wooden
washstand without drawers. *Spittoons* of hour-glass shape
are provided in the common rooms, placed on shelves at
a convenient height, and none in the corners. In the bedrooms it was found unsafe to use these, as patients amused
themselves in aiming at them from a distance, so that only

[1] *Die Bekämpfung und Heilung der Lungenschwindsucht*, Munich,
1897.

14

spitcups are used. there which must be taken up in the hand. Both kinds are partly filled with lysol solution. In addition to this, every patient receives on arrival a portable spitting flask as well as a thermometer; and the usual strict rules are adopted concerning random spitting or the use of handkerchiefs.

There are at present altogether 180 beds for patients. The increase in the numbers frequenting the *Krankenheim* is very remarkable, only twelve patients having been treated in 1894 and 510 in 1897. Patients agree to stay at least six weeks and usually remain twelve or thirteen. The payment is 26 marks per week in summer, 28 in winter. The majority are sent by the Insurance Companies and Benefit Societies, comparatively few paying directly for themselves. The men are under a house physician, with two or at times three male nurses (*Kraschnitzer diakonen*) ; the women have a separate house physician, sister and female nurse. In addition to this, one patient in each house is chosen by the rest to act as *Obmänn*. He has to take the temperatures every morning and evening (in the mouth), to see that rules are kept, and to report any breach of discipline or any complaint, which must be made in writing. The *Obmänner* collectively meet Dr. Weicker every Sunday, to report progress and make or receive suggestions.

The open-air treatment is carried out in the usual manner, excepting that the lounges are collapsible iron ones ;[1] and that, owing to the situation of the sanatorium, the patients have to use the public roads on their way to the woods. Patients use the same woods as those of the private sanatorium at Schmidtsdorf (see p. 157). The patients' beds are numbered, and each number entitles the occupier to three hours' rest in the fresh-air gallery, according to a pre-arranged list, thus preventing any clashing. Where longer rest is needed, the patient receives additional numbers.

[1] Described by Beaulavon, *Contrib. à l'étude du traitement de la Tuberculose pulmonaire dans les Sanatoria.* Paris, Bataille & Co.

In winter each patient receives two earthenware hot bottles, which he fills himself. The day begins early, breakfast being at 7 A.M. in summer, 7·30 in winter. The second breakfast is at 10 A.M., dinner at 1 P.M., another meal at 4, and supper at 7. The food is varied as much as possible, although difficulties are found owing to the daintiness of patients and the restricted dietary of the average German working class household. Patients of this class, it is said, will not touch fowl, game or fish, and often object to beef, as well as to a number of dishes commonly used in middle class households. Much butter and milk are used at the sanatorium, the proportions of albumen, fats and carbohydrates in the dietary being calculated for 121, 104, and 400 grms. respectively. The midday dinner consists of soup, one course of hot meat, two vegetables, and fresh or cooked fruit ; the supper of soup, one course of meat (hot or cold), and potatoes. The whole number of patients are brought together once a week to be weighed, and this occasions much friendly rivalry. The weight of the naked body is carefully estimated on entry and on leaving.

It has not been found practicable to let the patients do any work, as their capacity for work is very unequal, and yet each wishes to do like his neighbour ; moreover, in case of hæmoptysis or other complication, the physician is blamed for letting them work. Contrary to the experience of some other directors of sanatoria for the people, Dr. Weicker has not found the women any more easy to manage than the men. To occupy the patients, as well as to aid their recovery, systematic bathing exercises are prescribed ; courses of instruction in wood-carving, shorthand, etc., have also been begun and no evil results have so far been observed. Rubbing, baths and douches are prescribed in much the same way as at Dr. Weicker's private sanatorium.

Dr. Weicker and the senior resident medical officer are responsible for the whole conduct of the establishment, both in medical and in other matters.

The results of treatment have been very satisfactory. Out of 395 patients who left in 1897 with 30,245 days of treatment, 76·5 per cent. were restored to their full and 12 per cent. more to conditional working capacity.

For climate and other details, see also the description of Dr. Weicker's private sanatorium, p. 157 and also p. 150.

THE LOSLAU SANATORIUM

was opened in July, 1898, by the *Heilstätten-verein für Lungenkranke* for the district of Oppeln in Silesia, under the presidency of Regierungsrat Dr. Roth. It is a closed establishment exclusively for consumptives of the male sex belonging to the less wealthy classes, such as artisans, clerks, shopkeepers, teachers and the like. Arrangements have been made with the Silesian Insurance Society at Breslau to receive their clients in case of need.

The sanatorium stands at an altitude of 300 metres (984 feet), in 94 hectares (232 acres) of land, of which 91 hectares are pine woods. Touching the grounds is a large forest, the use of which is secured by contract. The building has a southerly aspect, and is sheltered from wind to the north, north-east and north-west. It is a somewhat plain-looking structure of wood and iron, with a long fresh-air gallery on the south side in front of the basement, above which are a ground floor and two upper floors. It has a large dining saloon and day room, balconies to the upper floors, and bedrooms to accommodate respectively one, two, three, and four patients each. The cubic space allowed in these amounts to 48 cubic metres (1695 cubic feet) per head, the ventilation being effected by open windows and air shafts. The walls are coated with enamel paint to a height of 2 metres (6 feet 7 inches), and above this are colour-washed. The floors are of cement, covered with linoleum. The building is warmed by warm-water pipes, and lighted by electricity. *Sputa* are received indoors into spittoons containing water; out of doors handkerchiefs are used and placed in tin boxes on entering the house. Linen is purified by boiling ; rooms by simply washing and airing. The douche is employed in suitable cases, but no cod-liver oil, alcohol or internal remedies.

There is room at present for eighty-seven, but there will be accommodation for 100 when a separate doctor's house is built.

None but early cases are received. Patients remain about three months, and pay 3 marks per diem.

The chief medical officer is Dr. Georg Liebe, to whom I am indebted for much useful information. It is stated that he will take charge of the Manebach Sanatorium when this is ready.

CHAPTER XXXIV.

SANATORIA FOR THE PEOPLE IN BRANDENBURG, MECKLENBURG, AND POMERANIA.

MOST of the sanatoria in this section are but little raised above the sea-level. Two of them (Malchow and Blankenfelde) might be regarded as homes rather than sanatoria for consumptives, as their position and arrangements are not very suitable for the fresh-air treatment. To this section belong the sanatorium at Grabowsee, the one being built at Belzig by the Berlin-Brandenburg Society, with its neighbour of the Bleichröder Stiftung, and the one which is being erected by the Berlin Insurance Co. at Beelitz. The Brandenburg Insurance Co. also proposes to erect a sanatorium for women at Kottbus, where the authorities have unanimously agreed to give the land in the municipal forest free of charge.[1] The Stettin Sanatorium Society will probably build a sanatorium at Höckendorf.[2]

THE BEELITZ SANATORIUM.

The Berlin *Invaliditäts und Alters Versicherungs Verein* has purchased 140 hectares (346 acres) of land at Beelitz, for the erection of sanatoria, convalescent homes, etc., at an estimated cost of six million marks. The land is cut through by the Wetzlar Railway, on one side of which will be a sanatorium for consumptives; on the other, similar establishments for those who are suffering from other ailments. There are to be twenty-six separate houses, with a total of 550 beds,[3] of which about 100 will be for the sanatorium.

[1] *Das Rothe Kreuz*, 1897, No. 20.   [2] *Heilst. Corr.*, Mar., 1898.
[3] *Das Rothe Kreuz*, Aug., 1898.

THE BELZIG SANATORIUM.

A large sanatorium for semi-necessitous patients is being erected under the auspices of the Berlin-Brandenburg Society, at Belzig in Brandenburg, about an hour and a half by rail from Berlin. The land acquired covers 1¾ hectares (4 acres). The building, of which the foundation stone was laid in August, 1898, is expected to be completed in about twelve or fifteen months, at an estimated cost of 540,000 marks. The main building is to be 150 metres long, with two wings of 100 metres each, and will accommodate eighty to ninety patients. Dr. Itzerott is to be the medical director.

THE BLEICHRÖDER SANATORIUM.

In the same grounds as the preceding a separate building for thirty beds is to be erected with funds left by the late banker, Herr v. Bleichröder, and this will be affiliated to the Belzig Sanatorium. The Bleichröder legacy was originally left for the treatment of consumptives by tuberculin ; but, in consequence of an appeal, the conversion to a sanatorium for hygienic treatment has been permitted by the authorities.[1]

THE BLANKENFELDE SANATORIUM

was originally a lying-in institution, which, being empty in 1892 at the time of the tuberculin craze, was converted into a women's sanatorium for consumptives. It has since then been gradually changed into a home for the reception of patients with various forms of chest disease. It is situated not far from Berlin, and has sixty beds. In 1896, 239 cases of consumption were received, and twenty-nine cases of bronchitis and emphysema, with 13,154 days of treatment, or an average of forty-nine days' stay. The total cost is 3·35 marks per day. Dr. Ellerhorst is the chief medical officer.[2]

[1] Liebe, *Hyg. Rundsch.*, 1st Nov., 1897.
[2] *Ibid.*, 1895, No. 17.

THE GRABOWSEE SANATORIUM

is situated in the midst of a pine wood on a hill near the Grabowsee, and is about 6 km. (nearly 4 miles) from Oranienburg, and about 18 miles north of Berlin. It was opened on 25th April, 1896, being the first sanatorium erected by the Red Cross Society, through the initiative of Geheimerat Dr. Gerhardt, who acts as consulting physician to the sanatorium, and examines intending visitors at the Charité Hospital in Berlin. The site covers 10 hectares (24 acres) of land, lying about 38 metres (125 feet) above the sea-level, on diluvial sandy soil, in a neighbourhood free from industrial smoke, noise and dust. The land has been obtained on a long lease at a yearly rental of 50 marks.

The sanatorium is intended exclusively for consumptive men who are likely to show decided improvement under treatment, and who are free from infectious or disgusting ailments. Three certificates are expected, one of which shows the name, age, place of residence, and name of insurance company or benefit society concerned; the second is a medical certificate of fitness; the third a guarantee as to payment. Patients are requested to bring a change of clothes, including underlinen, good socks and boots, comb and toothbrush, goloshes, umbrella, and if convenient sleeping jackets, slippers and the like. In case of need the institution will lend some of these articles, upon guarantee from the sick fund or benefit society. Bed and bedding are provided by the institution, which also supplies handkerchiefs, and washes the linen free of cost. The charge is 3 marks per day, payable for ten days in advance in the case of private patients, monthly in the case of insurance societies. The daily cost is said to be 2·47 marks. Although primarily intended for the working classes, Kuthy[1] states that shopkeepers are beginning to insure beds for themselves in the institution. A committee of ladies has been formed to look after the families of the poorer patients.

[1] *loc. cit.*

There are three free beds, one of which is the gift of the Empress.

The sanatorium consists of over two dozen buildings, most of which are light shelters (*Döckersche baracken*), with double paper walls, similar to those used in the army. When the sanatorium was first opened the patients were exclusively accommodated in these *baracken*, but as time went on a few more substantial structures were erected. There are at present some eighteen or more of these light shelters in use, one of which is utilised as a chapel, another as a storehouse, another as the medical officer's residence, another as a workshop, while the rest are summer sleeping quarters for the patients, or (with one side removed) are used as fresh-air shelters (*liegehallen*). They are exceedingly light and portable, being entirely of wood and linen-backed paper, everywhere covered with oil paint. The windows are threefold, the floors a wooden framework which may be covered with linoleum. Windows in the peak of the roof, and ventilators round the stove pipe, with open windows, ensure abundant ventilation. These *baracken* are made of various sizes. In this sanatorium they accommodate eight patients each, and are furnished with beds, long lounge chairs, a table, pedestals, washstand and stove. Ordinary tents were found to be far inferior to these structures.[1]

The remaining structures consist of an administrative block, two winter pavilions, laundry and disinfecting apparatus, and the gas making and storing apparatus. The administrative block, solidly built of timber, contains the kitchen, scullery, larder, serving room, dining saloon, office and library, quarters for one of the assistant medical officers in one wing, and for the sisters and female servants in the other. The kitchen contains a roaster, meat boiler and milk boiler over a water bath. The dining saloon is a large room, covering 127·19 square metres (1370 sq. ft.). In front of this building is a large fresh-air gallery. One

[1] Made by Christoph and Unmack, of Niesky, N. Lausitz.

of the winter pavilions, a single-floor structure of similar material, contains twenty-eight beds for patients. The other, with two floors, has fifty-two beds for patients, two day rooms (one covering an area of about 74 sq. metres = 794 sq. ft.), two consulting rooms, two lavatories, water-closets, douche and bath rooms, quarters of the sisters, a room for the bathman, and a common room. The bath and douche rooms are built out, and have a hot cistern over them heated by a steam-pipe. Over the main structure is a large cold-water cistern fed by steam power.

There are altogether 160 beds in the sanatorium, of which eighty can be used in the depth of winter. For the most part, two to eight persons sleep in each room, a few being single bedded. From 30 to 33 cubic metres (1060 to 1165 cub. ft.) air space is allowed per head. There are ventilating shafts in all the rooms, heated by hot pipes. In the winter pavilions, which are heated by low-pressure steam pipes, there are inlets through the walls near the radiators, and outlet shafts which pass under the corridors and converge to the common chimney over the chief heating apparatus. The administrative block and the *baracken* are heated by ordinary closed stoves with chimneys. The sanatorium is lighted with acetylene gas produced on the grounds. There is a steam laundry with disinfector on the grounds. In this building fresh air is brought into each of the steam radiators. The inspector lives on the first floor. The mortuary is next the laundry. There is a good water supply from a spring 16 ft. deep. The sewage, which is water-borne, passes through sewers with gullies at intervals to a filtering tank, and from this to underground pipes in an irrigation field. The sludge is used as manure. Kuthy describes earth closets and urinals as attached to each of the *baracken*, supplied with peat mould impregnated with 3 per cent. sulphuric acid. This system has apparently been superseded by water-closets.

For the open-air treatment, the fresh-air galleries attached to the permanent buildings, and those formed out of *baracken*, are used. Patients are allowed to fish, play croquet, and other games; an aviary and a number of rabbits also help to amuse them. In suitable cases they are permitted to do light work for remuneration. Walks may be taken outside the grounds with medical consent; but all resort to public houses is naturally forbidden. Most of the patients are treated with douches or friction with cold-water applications; many are taught to do respiratory exercises. Five *meals* a day are provided, including plenty of milk and butter. No cod-liver oil, alcoholic stimulants or " specifics " are given. The usual strict rules against spitting on the ground or into handkerchiefs are observed. The *sputa* are poured undisinfected down the water-closets. Linen is disinfected by steam ; the floors, walls and furniture are cleansed every week with a 1 per cent. solution of lysol. Temperatures are taken four times a day ; the patients are weighed every fortnight.

Dr. Brecke is the chief medical officer. He has several assistants. There are three sisters of the Red Cross in the sanatorium. The Brandenburg Insurance Society sends all its male consumptives to this establishment ; most of those from the Berlin Insurance Society also come here.

From April, 1896, to April, 1897, 264 patients were treated in this sanatorium, of whom 219 left the institution. Of these 14·1 per cent. were apparently cured, 64·4 per cent. more were decidedly improved : 2·3 per cent. died. Of those apparently cured, one stayed fifty days : twelve from fifty to ninety days; some as many as 210 to 240 days. Many are said to have left too soon.[1]

The sanatorium may be reached from one of two stations : Oranienbaum or Fichtengrund. The latter is a little over two miles' walk.

THE MALCHOW HOME FOR CONSUMPTIVES,

which is not a sanatorium so much as a hospital for the consumptive poor of Berlin, is situated at the village of

[1] Liebe, *Hyg. Rundsch.*, 1st Nov., 1897.

Malchow, about 80 miles north-west of Berlin, in the midst of meadows and irrigation fields, on sandy soil, 31 metres (102 feet) above the sea-level. For shelter against winds, which is very inadequate, it is dependent upon the trees in the park, which covers 3½ acres of land.

The building is a long low unpretentious brick building with two floors and three in the centre. On the ground floor are the common rooms, and next to them on the south side two verandahs, which are used as protected walks and fresh-air galleries, but are not sufficiently sheltered to be of much use in bad weather. In the upper floor, which is reached at either end by a wooden staircase from the garden, are the bedrooms and dormitories. Four of these contain two beds apiece, two contain twenty-four each, and two about sixteen ; eighty-six in all being admitted, and from 35 to 40 cubic metres (1236 to 1413 cubic feet) of air being provided per head. There is no central corridor, the rooms being placed immediately next one another. The heating is by low-pressure steam, the lighting by petroleum ; the ventilation by the windows, none of the rooms having chimneys. There are also a laundry and disinfection house, mortuary and water-filtering tank. Douche and shower baths for the patients will shortly be provided. The *sputa* are disinfected by means of hot soda solution ; the linen by boiling, the rooms by lysol. The sewage is carried onto an irrigation field. There are three trained nurses (Victoria Schwestern); and the institution is under the direction of a medical man (Dr. Reuter), who with the nurses has separate quarters in the old house belonging to the place.

Most of those admitted are consumptives, all of the male sex : but a few bronchitics are also admitted. In 1896, 555 cases of consumption were received, and 128 cases of bronchitis and emphysema, with a total of 26,551 days of treatment. All stages of consumption are admitted. Cod-liver oil, counter-irritants and so-called specifics are not much used, but a good deal of alcohol. According to Léon Petit, there is little or no systematic hygienic training, and very few facilities for carrying it out. The cubic space and means of ventilation are also somewhat inadequate. On these and similar grounds the institution has been somewhat severely criticised by Prof. v. Leyden, Léon Petit,. Beaulavon, and others.[1]

The home, which is not attached to any special insurance society, was opened 24th October, 1892. The daily charges are 2 marks. The daily cost is 3·16 marks. The building was originally intended for those who had successfully passed through the tuberculin treatment, but was thrown open in October, 1892, to consumptives without distinction, to the number of ninety-six, although the number admitted was afterwards reduced to eighty-six.

THE STETTIN SANATORIUM

is to be erected by the Stettin Sanatorium Society for eighty beds, on ground given by Stadtrat Dohrn on his property,. Höckendorf, near Finsterwalde. This is sheltered from the north and east.

A sum of £15,000 has been given by Dr. Karkutsch to the fund (*Heilst. Corr.*, March, 1898).

---

[1] Prof. v. Leyden, *La sollicitude des grandes villes pour les Tuberculeux*, Buda Pesth Congress, 7th Sept., 1894 ; Beaulavon, *Revue de la Tuberculose*,. Dec., 1896; Leon Petit, *Le Phtisique et son Traitement*.

CHAPTER XXXV.

THE sanatoria included in this section are the Albertsberg Sanatorium, the Plauen Sanatorium, and the Jonsdorf Convalescent Home.

THE ALBERTSBERG SANATORIUM,

or the *Heilstätte des Vereins für Begründung und Unterhaltung von Volksteilstätten im Königreiche Sachsen,* is situated at Auerbach, about twelve minutes' walk from Reiboldsgrün Sanatorium (described at p. 160), in the open country amidst hills and forests. The ground belonging to the sanatorium is but small (6½ acres), but the patients have the use of the surrounding State forest. The soil is granite, the elevation 700 metres (2297 feet). Woods protect it to the east and wooded hills to the north.

The sanatorium consists of a number of buildings united by covered corridors into the shape of a T (fig. 24), the centre consisting of two floors, while the rest is on one floor only. There are also several detached buildings for the dwelling of the chief medical officer and inspector, the kitchen block, ice house, coal house, stables and laundry. The main block has a southerly aspect, and has red roofs and white walls in the centre, flat " wood cement " roofs elsewhere. From the top of the T four pavilions project, containing dormitories for ten patients each. The space between them is occupied on each side by fresh-air galleries, in the centre by a large library and common- room 6 metres (19 ft. 8 in.) high,

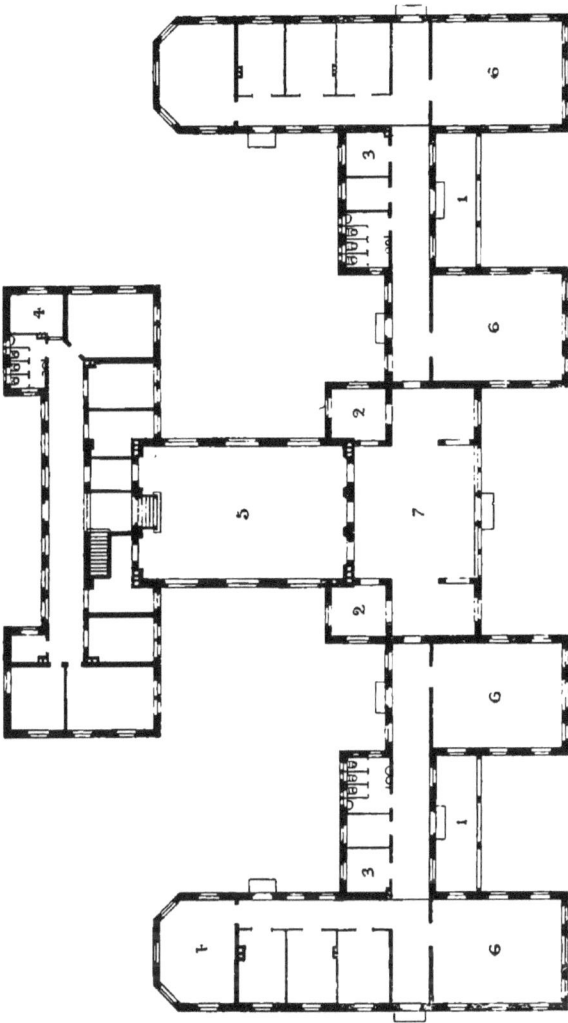

with two cloak rooms ; over these are double-bedded rooms, and a large balcony with side screens.   Behind, forming the central stem of the T, is the large dining saloon, which is 6 metres high, and capable of seating 100 persons.   Over it are eight single-bedded rooms.   The foot of the T is formed by a further expansion with bedrooms, etc., on two floors ; and behind this, forming a separate block, is the kitchen department.   Baths, closets and lavatories are everywhere on the north side.

There is altogether accommodation for 122 patients, some rooms receiving from one to four each, while a number sleep in dormitories ten together.   The cubic space allowed is 35 cubic metres (1236 cub. ft.) per head.   The ventilation is by open windows, the upper parts of which are valved ; also by ventilating valves and chimneys, which are often common to two rooms.   The heating is by closed stoves, but is to be altered ; the lighting by oil-gas with Auer incandescent mantles.   The waste water is carried through earthenware pipes to an irrigation patch.   The sewage is treated with dry peat mould.

None but consumptives are admitted, and only those in an early stage.   *Sputa* are poured into the water-closets without disinfection.   Linen is boiled with soap and water.   Rooms are daily cleansed with damp cloths.   The cold douche is used in all suitable cases; no cod-liver oil or specifics being given.   In addition to the fresh-air galleries next the house there is one in the woods.   For nursing there is one sister.   Dr. Gebser is the chief medical officer ; he was for five years house physician at Reiboldsgrün Sanatorium.

The institution was opened on 4th October, 1897.   It is connected with the Insurance Company of Saxony, the Railway Company of Saxony and the Ministry.   For the first of these fifty-five beds are reserved, for the others twenty-five each.   Patients' payments are at the rate of 3 marks per diem ; some pay only 2·50, a few being admitted free.   The building was estimated to cost 250,000 marks (£12,500).

Of this amount the Insurance Society advanced £4000 ; the central committee at Berlin £2500 ; a private gentleman gave £12,500 for a free bed ; the Railway Company pay £125 per annum ; and a private lady £150 per annum, besides other subscribers. The site was also given for a low price. An annual subvention is received from the Government.

THE PLAUEN SANATORIUM.

Another sanatorium for women is to be built in connection with the Albertsberg Sanatorium by the same society. The chosen site is 650 metres (2130 ft.) above the sea-level, in the midst of thick pine woods, sheltered by mountains to the north, east and west. It will be half an hour's walk from the men's sanatorium, but will have a common water supply and steam laundry. The building operations are to begin early in 1899 (*Heilst. Corr.*, Sept. 1, 1898).

THE JONSDORF CONVALESCENT HOME

is a "closed sanatorium" for patients with various ailments, including early cases of consumption, belonging to the union of *Südlausitzer Kranken Kassen.* It was opened in June, 1894, on land given by the town of Zittau, in Saxony. There are rooms for thirty-three patients, both sexes being accommodated in two separate portions of the building. There is central heating by low-pressure steam, and a private water supply.

The home is said to have cost 73,000 marks. It is under the care of Dr. Toop of Oybin. According to Hohe,[1] 93 per cent. of the patients who left the institution up to the end of 1896 were decidedly improved.

Jonsdorf is close to the frontier between Saxony and Bohemia, and is connected with Zittau by a narrow-gauge railway.

[1] *Die Bekämpfung und Heilung der Lungenschwindsucht*, Munich, 1897.

The Prinzessin Maria-Anna Heim.

According to Liebe[1] a sanatorium is being erected in Saxony by a local benevolent association for the benefit, of *Elb-Sandstein* workers.

[1] *Loc. cit.*

CHAPTER XXXVI.

IN this section are included the sanatoria of the St. Andreasberg district, and those at Stiege, Sülzhayn; Altenbrak, Königsberg, and Zellerfeld.

THE ST. ANDREASBERG SANATORIA.

There are two closed sanatoria for consumptives in the neighbourhood of this little town, the Oderberg Sanatorium and the Felixstift. Consumptives are also sent by a number of insurance companies to various houses in the town, under a common inspector, supervised by Drs. Hartung, Jacubasch, and Ladendorf. The town and climate, together with the provision for paying patients, are described at page 168.

THE ODERBERG SANATORIUM

is situated 640 metres (2100 ft.) above the sea-level, on the road from Andreasberg to Oderhaus, in the Harz Mountains, on the south side of the Oderberg. The district is scantily populated, and free from factories. The sanatorium is protected to the north by rising ground covered with woodland; and to the west, and partially to the east, by outstanding thickly wooded spurs of the Oderberg. To the south is falling ground, partly meadow and partly woodland.

The sanatorium consists of a main building, doctor's house, disinfection house, sundry stables and cowsheds, and two cottages for some of the servants. The main building,

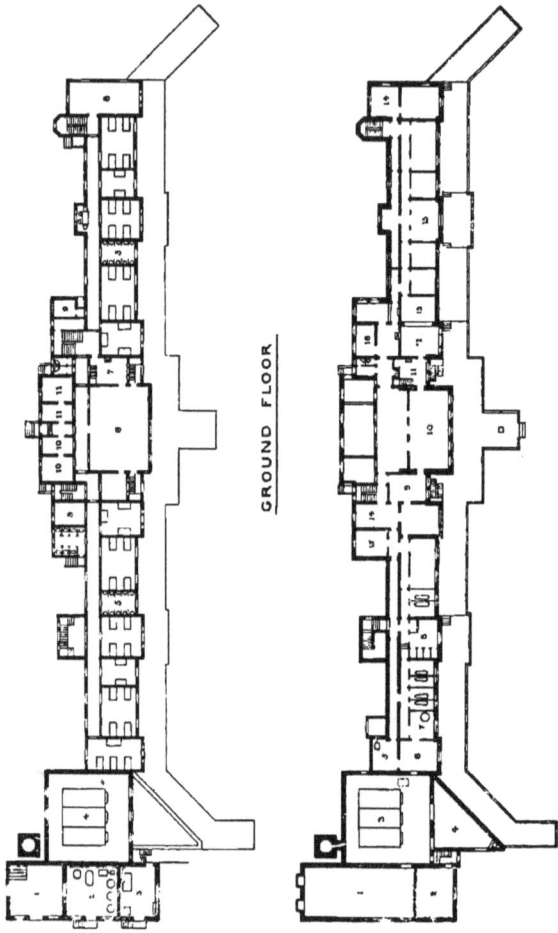

GROUND FLOOR

BASEMENT

FIG. 25.—ODERBERG SANATORIUM.

[Face page 227.

which has a due south front, and is built of wood on a solid
basement, consists of a central block and two wings, with
boiler and engine house at one end, all arranged in a long
line of 102 metres (335 ft.), about 100 yards from the main
road behind. Owing to the fall of the ground, the basement is above ground in front. It contains the kitchen,
scullery, bath and douche rooms, cellars and store rooms,
some of the ventilating machinery, and the workshop for
repairs, and has in front of it a fresh-air gallery with
diverging ends (see figs. 25, 26). The ground floor has in
the centre the common dining room, with the inspector's
consulting rooms behind, and on either side patients' bedrooms, which all have a southerly aspect. There are also in
each wing a lavatory and a nurse's room, and at the eastern
end a common room 28 × 13 metres (92 × 42½ ft.) in size.
The next floor has in the centre the inspector's quarters,
linen rooms, and others for the committee of management;
those in the wings resembling the rooms on the ground
floor. Above this is an attic floor, of which the centre is
occupied by bedrooms for the women servants, linen rooms,
etc., while the sides are like those on the lower floors,
excepting that in place of the common room there are four
isolation wards. These have no direct communication with
the rest of the bedrooms, although they are approached by
a passage from the common corridor. Each floor is served
by a longitudinal corridor at the back. For security in case
of fire, a massive wall with iron doors separates the western
wing from the rest of the building. Of the bedrooms,
seven are for one patient, fourteen for two, six for three
or four, fourteen for four each. A few more can be
accommodated in other rooms in case of need. The
average cubic space per head is 35 cubic metres (1236
cub. ft.); the average floor area, 10 square metres (108
sq. ft.), the height of the rooms being 3·80 metres (12½ ft.).
The walls are boarded and colour-washed; the floors of pinewood covered with linoleum, excepting in the kitchen and
scullery, where there are tessellated pavement and partially

tiled walls, and in the bath rooms, which have glazed tiles and a cement floor. The rooms have double windows. The furniture is of lacquered wood or iron, free from unnecessary dust-retaining irregularities. The dining tables are covered with American cloth, fastened with wooden fillets along the edges. The bedsteads are spring-bottom iron ones ; the mattresses are of horsehair in buttoned-up linen cases, with buttoned-up linen covers outside, and the same system is adopted for the pillows. The washstands and pedestals are of enamelled iron with glass tops. The lavatories on each floor have enamelled iron basins, with hot and cold water laid on, the plumbing resembling that of English lavatories. The water supply comes from springs in a meadow belonging to the institution, to the north of the main road, and is stored in three reservoirs.

The heating is by low-pressure steam, the lighting by electricity. Ventilating fans driven by electricity force a stream of air (which is warmed if necessary) into a long channel in the basement, which in turn supplies every room. In this way 80 cubic metres (2825 cubic ft.) per head can be supplied every hour. Inlets exist in every room near the ceiling; outlets near the floor ; these communicate with separate shafts and are covered with gratings. During the greater part of the year the incoming air comes through open windows ; in very severe weather the windows are closed and warm air forced in by the ventilating machinery.

Adjoining the western wing is a boiler and engine house, and next to it the accumulator room, with coal cellar and ashpit outside. The boiler and engine produce steam of 7 atm. pressure for the electromotor, reduced to 2 atm. for cooking and disinfection, and farther to $\frac{1}{10}$ atm. overpressure for warming the building and incoming air, and for heating the water for baths and scullery. The closets are movable automatic earth closets, in a partially built out and well-ventilated pavilion to the north of the basement corridor. The waste waters are disposed of by irrigation. Earth closet pails are also provided elsewhere for

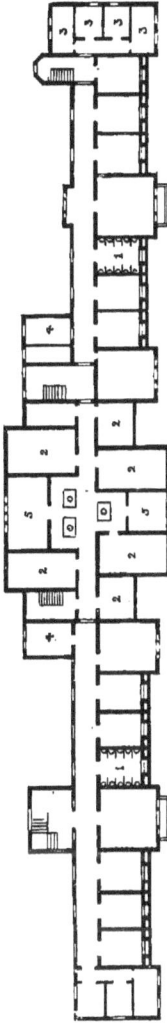

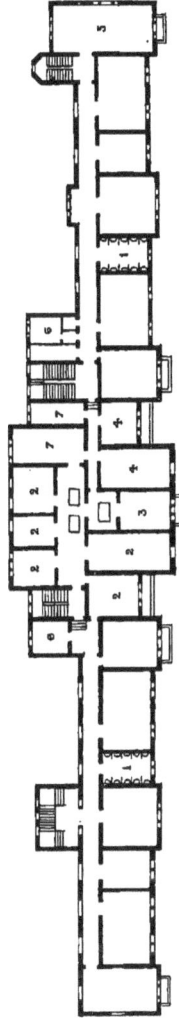

2ND FLOOR

1ST FLOOR

FIG. 26. ODERBERG SANATORIUM.

[Face page 228.

use during the night, and for the women servants, engine room staff, etc. The fresh-air gallery is $3\frac{1}{2}$ metres ($11\frac{1}{2}$ ft.) wide by 130 metres ($426\frac{1}{4}$ ft.) long, with wood cement roof, the height of which is from 2·75 to 4·6 metres (8 ft. 10 in. to 15 ft. 1 in.). It communicates directly with the bath rooms, and with the staircase to the dining saloon, and embraces with its diverging ends a sheltered gravelled recreation ground. It has places for sixty patients. The doctor's house is about 40 yards to the south-east, so placed as to command the fresh-air gallery. The disinfection house, which also contains a mortuary and *post-mortem* room, is about 50 feet westward from the engine house. The disinfector is double, with separate approaches for soiled and disinfected linen. Owing to the altitude, extra superheating is necessary to ensure disinfection. The stables, cowsheds, and houses for the servants, are still farther off (100 to 150 yards) to the west, the latter being near the road.

The staff, which is under an inspector, consists of twenty-five in all, and includes four male nurses (who also clean the rooms), a female cook and three kitchenmaids; a machine tender and his mate; a heater; a steward and his wife; coachman, and messenger. Some of the lighter duties, such as the cleansing of the spitcups, are performed by the patients, an overman being chosen for every section of the building (each containing about twenty patients). He is responsible for the tidiness of the rooms, the daily removal of closet pails, and the cleanliness of the closet of his own section, of which he keeps the key; for the regular use of the thermometer, the proper conduct of his own set of patients, extinguishing of lights, observance of rules concerning books, and the like.

The sanatorium, which was opened in August, 1897, under Dr. Liebe, is intended exclusively for consumptive men insured in the Hanse Insurance Society of Lubeck who are likely to recover their health or their earning capacity for a number of years. The average stay is three months; but this can, if necessary, be extended. The charges are defrayed by the insurance company. There is accommodation for 115, or in case of need 120. Since the 1st January, 1898, Dr. Ott has been the medical officer.

The treatment is that adopted in most of these institutions—rest or exercise in the open air, good ventilation, good food, and regular hours. There is no park belonging to the establishment, but there is plenty of open land around it; and patients use the fresh-air gallery and recreation ground, or the public pine woods near by. The douche is used after the weekly bath, not usually at other times. A very simple spitflask of blue glass with screw top is in use; only two spittoons existing in the whole establishment. The *sputa* are mixed with peat mould and burned. Linen is disinfected by steam; the rooms by thorough cleansing. In suitable cases gymnastics and respiratory exercises are prescribed. The *food* is the same as in other similar German institutions. It is served out for five at a time, the patients helping to distribute it. The milk is from a private dairy of twenty-three tuberculin-tested cows, but is all boiled before being used, in order to impress on the patients the importance of doing so at home. Beer is only given on Sundays. Specifics, cod-liver oil and counter-irritants are only used when specially indicated. Newspapers and political discussions are forbidden. Each patient is free to attend the religious services at St. Andreasberg with the doctor's sanction, or to abstain from so doing if he has conscientious scruples.

The nearest railway stations are at St. Andreasberg and Oderthal. St. Andreasberg is the terminus of a branch line from Scharzfeld.

A SANATORIUM FOR WOMEN

is likely to be opened in the same neighbourhood by the Hanse Insurance Company.

The Felixstift Sanatorium

is intended for ladies and gentlemen of small means rather than for the working classes, and owes its existence to the efforts of Dr. Ladendorf, the cost of erection having been largely defrayed by a legacy, and the cost of maintenance being guaranteed by a private society. Situated to the west of St. Andreasberg, about half an hour's walk from the town, it stands on a steep hillside, looking over meadows and a little stream to a pine-clad hill. It has a southerly aspect and is protected on all sides from cutting winds, the most distant view being to the south-west.

The building is a pretty structure in Swiss style, built of wood on a granite foundation, with red-tiled roof; and has accommodation for thirty-two patients at present, which number will be increased to forty when the place is completely furnished. It consists of a raised centre and two wings. In the western wing on the ground floor is a large dining saloon, with a small serving room to the north, a large window on the western side, and a large covered projecting balcony on the southern front. In the centre of the south side is the entrance hall with office and matron's and consulting rooms; and behind it the staircase. The rest of the ground floor and first floor is taken up with patients' rooms, some of which can be used as sitting rooms if necessary. The centre has a few attics in the peak of the roof. The basement, which is level with the ground in front, contains a light and cheerful kitchen in the left wing, with a small scullery, and near them a bath room and douche room, the larder, and various store rooms, together with the heating apparatus. The bedrooms, some of which have small balconies, are cheerful and airy, the walls painted, the wooden floor oiled and provided with a strip of carpet near each bed. The rooms are ventilated partly by means of open windows, which are in three parts independently movable; partly by warm-air inlets near the floor and outlets near the ceiling, which lead into

separate shafts. The bedsteads are of wood, provided with
feather beds and blankets. The heating is by low-pressure
steam pipes; the lighting by petroleum lamps. There are
six well-fitted water-closets and lavatories on the north side.
The sewage is carried to six clearing tanks, where it is
treated with alum and lime, being afterwards filtered
through sand on its way to the little stream. Four *meals*
a day are provided, in addition to a cup of tea in the
afternoon : the two breakfasts at 7 and 9 A.M., the dinner at
12 and the supper at 7 P.M. About 2 litres of milk per
head are used daily.

A simple and inexpensive *spitflask* is used in place of
Dettweiler's pattern. There are no spittoons. The resident
staff consists of the matron, three maids and a man. There
are no nurses. Dr. Ladendorf of St. Andreasberg is the
medical officer. Patients sleep two or three together in
one room, a few single-bedded rooms being also provided.

The charges in the common rooms are 23 marks per
week, including board, residence, medical treatment and
all other expenses. Where a single-bedded room and a
sitting room are engaged the charges are 4 to 5 marks per
diem.

There is a good road from St. Andreasberg. The institution was opened in August, 1898.

THE STIEGE SANATORIUM,

which is called Albrechtshaus, after the Prince Regent of
Saxony, has been recently erected (June, 1897) in the Harz
district by the Brunswick Sickness and Old Age Insurance
Company. It stands about 450 to 500 metres (1480 to 1640
feet) above the sea-level, near the narrow-gauge railway
from Gernroda to Hasselfelde, and a mile or two from the
little village of Stiege. The road to the sanatorium passes
through a wood, where the twin establishment, Marienheim,
is being built.

The sanatorium stands on a slope with meadow land in
front of it, and is surrounded on all sides with pine-clad

hills. It is a wooden structure on a granite foundation, with solidly built basement and dark-tiled roof, and is 68 metres long, shaped like the letter E. Behind the centre is the entrance, with the staircase and kitchen department around it. The kitchen, which is light and cheerful but rather small, and has tiled roof and floor, is supplied with hot cupboard, food lift, a central stove, etc., and has next to it a scullery and small store room. Above it on the first floor are the quarters of the house master, and above these a few garret rooms for women servants. Also in the centre, but on the south side, are the dining saloon on the ground floor, with a verandah in front of it, and two day rooms on the first floor, both of the latter provided with spacious balconies. Above these, in the peak of the roof, is a drying space. The lateral parts of the central block have a single row of rooms on the south side with a corridor behind. They comprise on the ground floor two bath rooms, two lavatories, the cook's bedroom, the office, consulting room, two patients' bedrooms each with four beds, and one with two. On the first floor are four rooms with four beds each, two with two beds apiece, a lavatory and room for the chaplain. In the wings on the ground floor are two large dormitories for ten patients each, which however owing to pressure on space have been made to accommodate fourteen apiece. In the lavatories each patient has his own basin and utensils. Under the large dormitories are the chapel in the eastern wing, a gymnasium in the western. The rest of the basement is occupied by heating apparatus and cellars. The patients' bedrooms have a cubic space per head of 27 to 28 cubic metres (9530 to 9880 cub. ft.). The windows are large, of the usual pattern with three sections, and a ventilator is present in each room. The walls are of rough plaster on the lower floor, oil-painted in the upper floor and the lavatories. The floors are of cement in the basement corridor, tiled in the ground floor corridor, elsewhere covered with linoleum. Most of the rooms are heated by closed stoves burning wood; but three rooms, including a bath and douche room, are heated by steam pipes.. The lighting is by petroleum. The closets are built out on the north side of the corridors. They are six in number and are earth closets, the user distributing peat mould with a spade. The waste waters are collected in a reservoir in the centre of the terrace in front of the building. In fine-dry weather they are allowed to irrigate the vegetable: garden ; in wet weather they are diverted into the wood to the east of the sanatorium. In front of the building is a fresh-air gallery, the roof and overhanging front of which are formed by a roller blind of sail cloth, the floor consisting of the pebble-covered terrace. A *Döckersche Baracke*[1] in the garden serves as a recreation pavilion. Behind the main building, across the courtyard, are stables, steam disinfector, etc.

Only men are admitted as patients, and not exclusively although chiefly consumptives. The patients do a little garden work and gymnastics, and are expected to ask for work after three weeks unless the doctor forbids it. They stay usually for thirteen weeks. The *food* consists of the usual five meals: beer being provided, and wine on Sundays.

During 1897 seventy-five patients were under treatment, with a total of 5213 days of treatment.

The staff number nine in all, including the house master. Dr. Köhler, who is medical officer, comes over from Hassel-felde two or three times a week.

THE MARIENHEIM,

a sanatorium for twenty female patients, is being built a little lower down on the same ground, and will probably be opened in the middle of 1899.

THE SÜLZHAYN SANATORIA.

In the southern part of the Harz, where the land begins to fall towards Göttingen and Nordhausen, is a little niche

[1] See p. 217.

in the mountain side overlooking the village of Sülzhayn. Densely wooded slopes rise up on three sides, forming a sort of amphitheatre, near the centre of which, on a precipitous rock, has been built the private sanatorium " Fernsicht," by the side of the rising pile of the Sülzhayn People's Sanatorium. The foundations of this imposing structure have been built up on massive stone pillars 18 metres (nearly 60 ft.) high, additional room being obtained by blasting the rock behind. On this artificial platform have been erected three buildings : the private sanatorium, 'the doctor's residence, and the people's sanatorium ; the first two already occupied, while the third is rapidly approaching completion. In such a situation there is almost perfect shelter against boisterous wind from the colder quarters ;. while a lovely view is obtained of the country to the south. There are nearly 14 hectares (34 acres) of woodland belonging to the institution. The soil is of porphyry and grauwacke (the latter used in building the sanatorium); the elevation about 500 metres (1600 ft.)[1] above the sea-level ; the climate that of the Harz Mountains generally—a dry, bracing, equable hill climate, with a fair amount of rain and snow at certain seasons.

THE SÜLZHAYN PEOPLE'S SANATORIUM

is planned somewhat like the one at Ruppertshain, with a concave front, a centre and two wings. The dining saloon is, however, placed behind the centre on the first floor, and communicates by a bridge with the kitchen block still farther to the north. The patients' bedrooms, which are placed in a single row on the south side, on three separate floors, contain from one to four beds apiece, with an allowance of 42 cubic metres (1480 cub. ft.) per head, and a height of 4 metres (13 ft.). The floors are made of *torgament*, a patent material consisting of wood shavings incorporated with cement, which is capable of uniting with stone,

---

[1] According to Liebe ; Dr. Kremser gave me the height as about 1450 ft.

iron or woodwork, and is warmer to the feet than ordinary cement. The walls are of colour-washed plaster of Paris, the doors of wood without projections and very carefully fitted. The windows are large, but have not been carried up to the ceiling. Throughout the sanatorium all angles are rounded and surfaces smooth. Running in front of the centre block on the basement, which on the south side is on the ground level, is a corridor 4 metres (13 ft.) wide, which can be heated, and serves as a promenade in bad weather. It has very large windows which open out on to a covered strip of garden, and boot rooms on either side. The corresponding parts on the ground floor form a large day room and a fresh-air gallery 5 metres (over 16 ft.) wide, to be used for rest in the open air, and protected by the balcony overhead. In both the centre and the wings balconies are placed in front of the bedrooms. The lavatories, water-closets, rooms for the nurses, and staircases are at the back of each wing beyond the corridor. There are two water-closets in each wing on each floor. The water-closet basins cover themselves automatically. Next the water-closets are rooms for disposing of the *sputa*, etc. In one compartment is a boiler for boiling the sputum, after which it goes down a special soil pipe into the drain. In another compartment the spitcups are boiled and the chamber pots cleaned. The common bath rooms and lavatories, which are also to the north, have terrazzo floors covered with parallel wooden laths. Each patient will have his own china basin fixed to the wall, with a box for toothbrush, etc. ; on one side of the lavatory there is only cold water ; on the other hot and cold. The douche rooms, with doctor's room and dressing room, are in the wings on the ground floor, and have the usual arrangement of taps and thermometers to indicate the temperature of the douche. Underneath are the boiler for disinfection of sputum, the inhalation room, and some cellars. There are also a dark room for photography and a library. Each wing has also a lavatory for the nurses, and a small store room ; and on the lower floors a common room for the patients.

Behind the centre is the administrative block. This has in the basement the accumulator room and coal cellar, with the ice cellar on one side and the three heating furnaces on the other. Above these come the laboratory and consulting room, the porter's room and office, on each side of the entrance. On the next floor is the large dining saloon, with a room for the sisters on one side and a couple of committee rooms on the other. This communicates with the kitchen block by means of a serving room on the bridge. The kitchen block, which has its entrance behind, contains in the basement the larders, store rooms, cellars and engine room. There is also a passage under the road to the main block. On the ground floor is the laundry department with disinfecting apparatus and linen rooms on either side. The disinfector has two approaches; the linen rooms are for storing and mending respectively. The laundry has an electric mangle, and rolling and ironing rooms. An electric lift carries the washed linen up to the drying loft. Above the laundry department is the kitchen department, with scullery, crockery room, vegetable kitchen, and dining room for the servants. The kitchen is separated from the serving room by a passage, and has a special ventilating shaft. On the next floor are the quarters of the engineer and his wife, together with a separate portion for the women servants, and a drying loft. The kitchen department has its own staircase, douche and bath rooms, and water-closets.

The sisters are lodged in separate quarters in the roof of the main building. The sanatorium is heated by low-pressure steam pipes, and by closed stoves in addition. It is lighted by electricity. There is an abundant supply of good water, which works a turbine, and has a pressure of 5 or 6 atmospheres. The staircases are of stone, let into the thickness of the walls, without other support.

There is altogether accommodation for 100 patients, but

twenty more could be lodged in the attic floor if necessary. Only men are admitted ; the sanatorium having been built by the N. German *Knappschafts Pensions Kasse* in Halle-on-Saale, chiefly for the miners of the district.

Dr. Kremser is the chief medical officer. He has under him a matron and female nursing staff. The assistant medical officer is also appointed on his recommendation, subject to the assent of the matron. All the other officials are directly appointed by the chief medical officer. They consist[1] of an engineer, female cook, two kitchen maids, two male nurses, messenger and night watchman. No inspector will be appointed. The male nurses clean the spitcups. The building is to be cleaned daily by women from the village.

During the building operations the poor patients have been lodged in the private sanatorium "Fernsicht," and before this was built an open colony existed in the village.

THE FERNSICHT SANATORIUM

lies to the west of the miners' sanatorium, and is attached to Dr. Kremser's residence, which is between them. Owing to the rapid fall of the ground it has a deep basement which is above ground on the south side, whereas the doctor's villa has one floor less. There is accommodation for ten patients on the ground and first floors, all being to the south and provided with balconies. Between the two buildings is a deep fresh-air gallery on each of these floors. The basement contains in front a reading room and billiard room, and behind a bath room and consulting room. Over the latter on the two upper floors are bath rooms and lavatories, together with rooms for the attendants. The south-eastern corner is occupied by the furnace room, above which is the kitchen, and above this the dining saloon, with a large winter garden in front of it (fig. 27).

The charges will be probably 6 marks per diem.

---

[1] Liebe, *loc. cit.*

FIG. 27.—DR. KREMSER'S SANATORIUM "FERNSICHT,"
SÜLZHAYN IN THE HARZ.

[*Face page* 238.

ANOTHER SANATORIUM

under the charge of Dr. Kremser's brother-in-law exists in the village of Sülzhayn, about 150 to 200 feet below the Sülzhayn miners' sanatorium. ' There is accommodation for nine patients, who pay from 4 to 6 marks per diem. Additional beds can also be obtained in other houses in the village.

ALTENBRAK SANATORIUM

is situated in the Bodenthal, which some consider to be the finest part of the Lower Harz district. It consists of the Villa Emma and other houses in the village, where patients are received from various insurance societies, and is managed by an assistant of Dr. Pintschovius of Ketzin, who also owns a small private sanatorium in the same neighbourhood (see p. 167).

THE KÖNIGSBERG CONVALESCENT HOME

is situated about $3\frac{1}{2}$ kilom. (2 miles) from Goslar, in the northern part of the Harz district, in the midst of beautiful pine woods, and is 450 metres (1470 ft.) above the sea-level. It was originally a private villa, but was converted into a sanatorium by the addition of a large dining saloon. Opened in May, 1895, by the Hanover Sickness Insurance Company, purely as a convalescent home for men, it was after a time devoted chiefly to consumptives, owing to the number of other convalescents attached to the insurance company being insufficient to keep it full. The building has a south-east aspect, and has a meadow in front of it with a rapid downward slope. Across the valley in front is a wooded hill; on each side of the meadow and the sanatorium are pine woods. The ground also falls to the south-west, and rises to the north. Owing to the height of surrounding hills, and the neighbouring pine woods, there is good shelter against cutting winds.

The building is of wood, in the Swiss style, with a turret in the middle. The chief entrance is at the south-west end, which leads into a corridor running parallel to the front, with rooms on either side. At the opposite (north-east) end is the new wing, containing the kitchen and dining saloon. From the latter a door leads on to a broad verandah, and thence to the garden. The bedrooms for patients are on the ground floor and first floor. They are placed on all sides of the building excepting the north-east, and hold from two to four beds each. They are light and cheerful, with linoleum-covered floors, washable paper on the walls, varnished wooden wainscots, large French windows (mostly with balconies), and plain, clean-looking furniture. The bed-rooms on the second floor are for the staff. There is a reception room on the ground floor near the entrance. Hot and cold water are laid on to every floor. There are three earth closets, but only one bath room. The basement contains heating apparatus, engine and dynamo room, small accumulator room, four cellars, and a larder; and in the added wing a cheerful kitchen with tiled walls and floor, and windows on three sides. Near this are a disin-fecting room, laundry, and stables, and over these a drying space and the quarters of the farm manager. These are all separated from the rest of the house by a narrow passage. The lighting is by electricity, the heating by low-pressure steam. A two horse-power engine supplies a 36-celled accumulator; these suffice both for heating and lighting, and for pumping water from the meadow to a cistern under the roof.

There is accommodation in the house for thirty-six patients, and fourteen more are lodged in a *Döckersche Baracke* [1] in the meadow. Ninety per cent. of those admitted are consumptives. The usual five *meals* a day are provided; the food is said to be excellent. Beer and Bordeaux wine are allowed. Patients do light work for wages if the doctor permits, and also the lighter house-work. *Spitcups* are disinfected with corrosive ublimate.

The institution is managed by a sister of the Hermitten-

[1] See p. 217.

stift in Hanover, and is under the care of Dr. Andrae of Goslar, who comes over twice or three times a week. During 1897, 178 patients left the institution, of whom 107 were tuberculous. Eighty of the latter (or 75 per cent.) regained their full working capacity, and 17 per cent. were apparently cured.[1] The daily cost is said to be 2·23 marks, or, with interest on capital, 2·71 marks.[2]

THE ZELLERFELD SANATORIUM,

which was opened by the Hanoverian Sickness and Old Age Insurance Company in August, 1898, for the reception of women, is situated in one of the most elevated and open parts of the Harz district. Leaving Goslar by rail the train goes panting and puffing up a steep incline through a winding river valley skirted by dense pine woods, until it reaches the bare and open mining country near Clausthal. Adjoining this important mining village is the more purely residential village of Zellerfeld, which climbs the hillside on the way to an elevated plateau consisting chiefly of meadows and cultivated land. About a mile from the village, near a small lake and stream, is a pine wood, by the side of which the sanatorium stands. Originally a restaurant and brewery going by the name of " Zur Erbprinzentanne," it is next the main road opposite a village inn. This circumstance is probably less disadvantageous in a women's sanatorium than would be the case in one for men. Altogether 5½ hectares (13 acres), mostly woodland, belong to the institution, which with the exception of the above-mentioned inn has no other buildings near it. The elevation is 550 metres (1800 feet) above the sea-level, the soil being of sand or rock.

The sanatorium consists of an old wooden structure and a brick-built house to which have been added a new one of timber and plaster on a brick foundation, and some stables. These form three sides of a square, and surround a courtyard which will be partly covered with grass, partly paved with

[1] *Report of the Inv. and A. Versicherungs Anstalt of Hanover for* 1897.
[2] Liebe, *loc. cit.*

16

granite blocks—asphalt not standing the severe cold met with in this district in winter. Across the road to the north-east is a sheltering line of pine trees, and beyond them a sudden fall to the stream ; behind the sanatorium, and to the sides, a thick belt of trees. High pine-clad hills may be seen on the horizon in most directions ; but the pine trees around the building form the only efficient wind screen.

Each part of the sanatorium consists of a basement, mostly occupied by cellars, and of a ground floor, first floor, and second floor under the roof. On the right of the court-yard at the north-west end is the original wooden building, the ground floor of which is partly occupied by the con-sulting room and administrative rooms. At the junction between this building and the centre block are the quarters of the nursing staff. The centre block has a large day room and the dining saloon on the ground floor, and behind these a wide verandah which also extends along the north-west end. The corridor serving these rooms is next the court-yard, and leads to the closets, of which there are two on this floor, one for the staff and one for the patients. The kitchen, which is also on the ground floor, has a central stove and tiled walls. At the extreme end of the centre block are two bedrooms separated by a short passage from the rest for cases requiring isolation. On the next floor are most of the bedrooms ; there are nine of these, accommo-dating from two to five apiece, or a total of forty. The walls are painted, floors covered with linoleum. The bed-steads are of iron, the mattresses partly stuffed with vege-table fibre. The washstands are of enamelled iron with drawers and glass tops. The pedestals are also of open iron work with glass tops. There is one bath room with two fixed baths. The heating is by means of closed stoves. The staircases are of cement. The further wing is com-posed of stables and cowsheds, together with the farm manager's house.

The sanatorium is connected with the telephone system. Dr. Plümecke is the medical officer.

CHAPTER XXXVII.

SANATORIA FOR THE PEOPLE IN THURINGIA.

THESE include the colony on the Harth, the new Berka Sanatorium, and the Manebach Sanatorium. An open colony exists at Gross Tabarz, near Friedrichroda, on an elevated hill-girt plain 1300 feet above the sea-level. At Erfurt (which although not in Thuringia is surrounded by it) a sanatorium is to be built by the Johanniter Knights out of the proceeds of a legacy.[1]

THE COLONY IN THE HARTH FOREST,

which is a few miles from Bad Berka, near Weimar, in the midst of beautiful pine woods, was opened on 27th May, 1896, and consists of Schloss Rodberg, with twenty to twenty-five beds, the Restaurant Sophienhöhe, which has eight to ten beds for summer use, and a number of huts in the woods which are also only habitable during the warmer parts of the year. The colony is in charge of a nursing sister; and consists mainly of patients from the Thuringia Insurance Company. Dr. Münsel of Weimar comes over frequently to supervise the treatment.

Of eighty-three patients who left the institution in 1896 (six months) seventy were phthisical or suspected to be so. Of this number seven presented no traces of disease on leaving; twenty-five showed traces, but were fully capable of work; twenty-five more were capable of working, but were less likely to remain so; eleven were only capable of light work, and in twelve no good result was obtained. Three did not complete their treatment.[2]

[1] *Brit. Med. Journ.*, 21st May, 1898.   [2] Liebe, *loc. cit.*

THE BERKA SANATORIUM,

which is being built by the Red Cross Society of Weimar, and is rapidly approaching its completion, will probably take the place of the above-mentioned colony in the Harth Forest. The site is on a picturesque headland in the midst of the pine woods, overlooking the little town of Tannroda with its ruined castle, and the valley of the Ilm. The soil is a beautiful red sandstone, which has been utilised in building the institution.

The building, which has a southerly aspect and is well protected to the north and east, is in the form of a central block with wings projecting back and front. It will accommodate men only, and will have about eighty beds, each room containing from two to four beds and having a southerly aspect. The larger bedrooms are in the wings in front. Behind the east wing are the bath and douche rooms ; behind the west wing the sisters' quarters. There are altogether two bath rooms and a large douche room. In the centre of the middle block is a large and lofty dining saloon, and next it a day room. The kitchen is behind, separated from the dining saloon by the corridor. The windows are all large, but do not reach the floor or the ceiling. The walls are lime-washed ; the floors of wood in the bedrooms with fillets to fit into the angles ; in the corridors they are of *torgament* (see p. 225). There will be central steam heating, with two heaters under the kitchen, and long, large steam-pipes under the windows. A ventilation shaft containing a central steam-pipe is placed in the middle of the building. The place will be lit by electricity. There will be eight water-closets with a good water flush. Behind the main block is a courtyard, on the far side of which are being erected stables, disinfecting rooms, etc.

Dr. Münsel of Weimar is to be the resident medical officer. The institution will be utilised by the insurance association of Sachsen-Anhalt in Merseburg, as well as by the Thuringian Insurance Company.

THE MANEBACH SANATORIUM.

A sanatorium is also being built for women at Manebach, near Ilmenau, 525 metres (1720 ft.) above the sea-level. There is already a recently built Curhaus, which, with an adjoining house, will accommodate forty patients. The new building is to be still larger ; and when it is completed the Curhaus will be devoted to a different class of society.[1]

[1] Liebe, *loc. cit.*

CHAPTER XXXVIII.

IN addition to the open colonies at Wissen-on-the-Sieg and Uckerath, near Oberpleis, and those of Lippspringe and Salzuflen in Lippe Dettmold, there is a sanatorium at Rehburg and another at Altena. Another is projected at Hagen in Westphalia, which is to be called the *Kaiser Wilhelm* Sanatorium, and for which 270,000 marks have been subscribed.[1] There are other projects for the establishment of sanatoria for Aix-la-Chapelle, Oldenburg, Altona, and Hamburg respectively. For the latter a legacy has been left of 250,000 marks ; and the State of Hamburg will contribute 60,000 marks per annum.[2] A sanatorium is also projected at Lippspringe for the consumptives of Minden.[3] There is already a Johanniter Hospiz there.

THE REHBURG BREMEN SANATORIUM

was opened on 1st June, 1893, by the *Bremen Heilstätten-Verein für bedürftige Lungenkranke* at Bad Rehburg, a little to the north of the Royal Bathing Establishment. It was the third of those opened in Germany for the poor, being preceded by the one at Malchow and the small one at Falkenstein, now replaced by the Ruppertshain Sanatorium. Originally containing twenty-four beds, it was enlarged in 1894 by the addition of wings, raising the number to thirty,

[1] *Heilst. Corr.*, July, 1898.

[2] *Deutsches Central Komité zur Err. von Heilst. f. Lungenkr.* Ann. Rep. for 1897.

[3] *Das Rothe Kreuz*, 1st Sept., 1898.

(246)

FIG. 27*.—THE BREMEN SANATORIUM, BAD REHBURG.

twelve for women and eighteen for men. It is under the
medical care of Sanitätsrat Dr. Michaelis, the owner of the
private sanatorium, who with his assistant attends the
Bremen Sanatorium gratuitously. It stands in a very
sheltered position on the hillside, and owing to the slope
of the ground the western half of the building has been
built on a slightly higher level than the eastern. Close to
it are corn fields ; a small garden lies to the south, and
densely wooded hills protect it to the north and west, and
the high beech trees of Bad Rehburg to the east.

The building consists of a centre and two wings, which
form a nearly unbroken line facing south. In the centre
are a dining room, day room, doctor's room and matron's
room, and in the projecting bow a room with four beds.
The east wing, which is for men, contains two large rooms
with four beds each, and two smaller with two beds each,
and under the roof another room with two beds. The west
wing, for women, has two rooms with four beds each, and
two rooms with two each. The rooms are 4 metres (13 ft.)
high, the larger being 7 × 9 metres (23 × 29½ ft.), the smaller
3½ × 7 metres (11½ × 23 ft.). The walls are colour-washed ;
and in the dining room there is also linoleum to the height
of the shoulders. Floors are everywhere covered with
linoleum, which is daily cleansed with damp cloths. The
place was beautifully clean when I visited it, simple, homely
and practical. The heating is by means of close (Lönhold)
stoves ; the lighting with petroleum. The furniture is
simple ; bedsteads of iron, with spring and horsehair
mattresses. Beds are disinfected when necessary by steam
heat in the public disinfecting oven. The rooms are simply
cleansed and aired. *Sputa* are poured down the water-
closets. Patients are instructed to use spittoons.

The sanatorium is managed by a matron (a sister of the
Red Cross), who does what nursing, rubbing, etc., are needed.
She has under her a cook and kitchen maid, housemaid, and
a man.

Treatment is by hygienic methods. Five *meals* a day

are provided, the mid-day dinner being a substantial meal
with several courses. Beer is provided at supper. Patients
who are medically fit help in the lighter household duties,
such as making the beds, cleaning their own boots, and
cleaning the spitcups.

They pay 2 marks per day if connected with the Bremen
insurance societies; 3 marks if from other towns. A few
gratuitous beds are provided whenever the finances permit.
No charge is made for medical attendance, drugs, or in
some. cases for personal washing. There is a fund in Bad
Rehburg for the assistance of poor people who come
for treatment, baths and other means being gratuitously
provided.

Up to the end of 1896, 334 patients were treated at the
sanatorium, with 27,287 days of treatment, or an average
of 81½ days. Only 297 were certainly consumptive, 43½
per cent. being seriously affected. Excluding the doubtful
cases, 85 per cent. improved in general health, 8 per cent.
lost ground. The lung condition improved in 21 per cent.,
grew worse in 13½ per cent. In 1897, 106 patients left the
institution. Of these 44·3 per cent. appeared to be permanently able to work ; 32·1 per cent. were thoroughly
capable of working, but with less certainty of a lasting
result.

The climate of Bad Rehburg is described at p. 171.

THE HANOVER SANATORIUM SOCIETY,

which was originally connected with the Bremen Society,
also sends patients every year to various houses in the
village of Bad Rehburg as well as to the convalescent
home at Königsberg and the Zellerfeld Sanatorium. These
are described at pp. 239, 241.

THE ALTENA SANATORIUM

for the district of Altena, in Westphalia, was opened on
1st August, 1898, in the open country between the villages
of Hellersen and Brünninghausen, about 4 kilometres (2¼

miles) from the town of Lüdenscheid. It stands 420 metres
(1380 ft.) above the sea-level, in about 38½ hectares (nearly
95 acres) of land, on the south side of a hill which is wooded
with oaks and pines and sheltered to the north-east by more
distant hills. The grounds are partly wooded, but in front
of the building consist of a large open grass plot, with
gardens and little ponds.

The sanatorium buildings consist of a main block, an
administrative block, doctor's villa, and open-air galleries.
The main block has a basement, ground floor, and three
upper floors, all the patients' rooms being to the south.
Behind it is the administrative block, which is united to
it by a covered bridge, and contains the kitchen, scullery,
store rooms, dining saloon and nurses' dining room, and
quarters for the second medical officer and the female
staff. The open-air galleries are not placed in front of
the main block, but form a curve on each side of it,
the western gallery leading to the chief medical officer's
residence. It is intended to build a chapel at some future
time.

There is accommodation for 100 male patients. The
bedrooms are of various sizes, and contain from one to
eight beds each. They are furnished with iron bedsteads,
steel and horsehair mattresses, iron pedestals with glass
plate, and a clothes cupboard for each patient. The lavatories are separate from the bedrooms, and have water laid
on over each fixed washbasin, and shelves above for washing utensils. There is also a large bath room with three
fixed baths and a dressing bench of xylolith plates. Next
to this is a douche room, with various douche and spray
apparatus, and a compartment provided with two water
cocks and a thermometer for the medical officer to regulate
the temperature of the douche. Another bath room is provided for the staff. The water-closets have porcelain basins
with oaken seats and automatic waterflush. The urinals
also have automatic water supply; their walls are of
polished granite, the floors of terrazzo. The sewage and

waste water are carried into settling tanks, and purified by filtration. The water supply, which is said to be good and abundant, comes from special springs in the neighbouring mountain, and is pumped by a benzine motor from a reservoir near the administrative block to another in the woods 4 metres higher than the roof. There are two day rooms in addition to the dining saloon. Every room is provided with ventilation shaft, and with special ventilation for winter. The building is heated by low-pressure steam. It is at present lighted with petroleum, which will, however, probably be replaced by electric lighting. There are electric bells in every room, with pushes near each bed. All the patients' rooms and the corridors are covered with linoleum; the bath rooms being tiled. A special room for boot cleaning is on the ground floor. The kitchen has a tiled floor, and is provided with two steam boilers and a roaster, with a food lift to the serving room above. The dining saloon is 13 × 7 metres (42½ × 23 ft.) and 6 metres (19 ft. 8 in.) high. There are a steam disinfector with two approaches, and a laundry with ironing and drying rooms. The washing is done by machinery, driven by a benzine motor. The *sputa* are disinfected with lysol solution.

The whole staff is under the control of the medical superintendent, who will be aided by a second medical officer. There are at present three Red Cross sisters from Cologne, one of whom acts as housekeeper, another as secretary, while the third sees to the nursing. When more patients are under treatment there will be one or two more nurses. There are also a bathman, a porter, a female cook with two kitchenmaids, and two women for the laundry work. The medical officer is Dr. Stauffer, to whom I am indebted for most of these details.

None are admitted but consumptives who have a reasonable prospect of recovery. There is an arrangement at present with the Westphalian Old Age and Sickness Insurance Company, but patients will probably also be sent from the Hanover Insurance Company and from the Rhine province.

Patients pay 3½ marks per diem, or 5 in a single-bedded room, and stay at least twelve weeks, unless otherwise determined by the medical officer.

Lüdenscheid station may be reached *via* Hagen and Brügge. The sanatorium is connected with the telephone system.

OTHER INSTITUTIONS AT ALTENA.

There is also at Altena an "Isolation Establishment" for consumptives in connection with the Johanniter Hospital in the same town. Treatment lasts from one to three months ; the payment is at the rate of 3 marks per day.

Two or more funds exist for the support of the families of patients who are under treatment in sanatoria, and who are in reduced circumstances. One of these has a capital of 140,000 marks (£7000), and will also be applied to the improvement of patients' dwellings. Another of 10,000 marks has just been started in memory of the late Count Bismarck, who was an honorary citizen of Lüdenscheid.[1]

[1] *Heilst. Corr.*, July, 1898.

CHAPTER XXXIX.

SANATORIA FOR THE PEOPLE IN THE RHINE DISTRICT.

EXCEPTING the Ruppertshaiu Sanatorium in the Taunus:
no large sanatorium for the poorer classes has yet been built.
in this district, although several are projected. A society
has been formed for this purpose with branches at Duisberg,.
Essen city and district, and Ruhrort.[1] Another union of
local societies, which is called Gesellschaft Gemeinwohl,.
and includes Barmen, Dusseldorf city and district, Elberfeld, Mettmann, Solingen town and district, Lennep and
Remscheid, has made greater progress. A Dusseldorf firm.
has promised 20,000 marks, and an Elberfeld banking firm.
10,000 marks ; and it is hoped that building operations may
be begun in the autumn of this year.[2] Barmen at present.
sends its consumptives to an open health resort at Godes-
berg near Bonn. The insurance company for the Rhine.
province will probably support the new sanatorium.
Cologne has a *Verein zur verpflegung Genesender*, with
five convalescent homes, and sends its consumptives to two.
of these (at Wissen an der Sieg for men, and at Uckerath
near Oberpleis for women); but proposes to build a closed
sanatorium as well.[3] There are a few beds for consumptives.
at the Philomenen Hospiz at Honnef. On the opposite bank,
Dr. Achtermann receives patients of the poorer classes.
into the " second-class section " of his sanatorium at Laub--
bach (see p. 189), but this establishment was not originally
built for such a purpose. A society has also been started at,

---

[1] *Heilst. Corr.*, July, 1898. [2] *Das Rothe Kreuz*, 1st Sept., 1898.
[3] Liebe, *loc. cit.*

FIG. 28.—VIEW OF RUPPERTSHAIN SANATORIUM.

FIRST & SECOND FLOORS

FIG. 30.—RUPPERTSHAIN SANATORIUM.

1. Dining Saloons.   2. Large Dormitories.   3. Single-bedded Rooms.   4. Larger Bedrooms.
5. Nurses' Rooms.   6. Small Kitchen.   B = Bathrooms.

[Face page 253.

Wiesbaden for the erection of a sanatorium of fifty beds for the lower middle classes of Wiesbaden, the Rheingau and the Unter Taunus.[1] At Hanau 30,000 marks were subscribed for a sanatorium fund, to commemorate the 300th anniversary of the town ; and half of the interest will be used to send patients to existing sanatoria, the rest being added to the principal.[2] There is also said to be a sanatorium for 120 beds in course of erection at Höchst am Rhein. The districts bordering on the lower part of the Rhine will be included in another chapter.

THE RUPPERTSHAIN SANATORIUM

is the successor of a small establishment at Falkenstein, founded by Dr. Dettweiler for the consumptive poor in 1892 with the help of the Frankfort Convalescent Association. This earlier sanatorium, which was merely an ordinary village house, was replaced in 1895 by a specially constructed building near Königstein in the Taunus Mountains, about forty minutes' drive from the older institution. The new sanatorium is at the head of a wide valley, with a south-easterly aspect, and is sheltered on three sides by mountains, for the most part covered with trees of various kinds. The soil, as at Falkenstein, consists of slate, gneiss, and porphyry ; the climate is that of the Taunus district generally. The elevation is 400 metres (1300 ft.) above the sea-level. The grounds of 4½ hectares (11 acres) are partly wooded with newly planted trees, partly meadow and cultivated land.

The sanatorium is built near the road in the form of part of a circle, with thickened ends (fig. 28), and has a basement, ground floor, first and second floors, and an attic floor within a steeply pitched and lofty roof. Projecting from the ends are two fresh-air galleries on the basement level, leading on to a terrace in front, with a view over the Main valley as far as the Odenwald. Behind the galleries on each side is a built-out pavilion, connected with the main

---

[1] *Heilst. Corr.*, July, 1898.   [2] *Ibid.*

block by a short corridor. The western pavilion contains the kitchen department; the eastern, the stables and wash-house.

The basement (fig. 29) of the main block is occupied by cellars, heating apparatus, and bath and douche rooms on either side. The ends, which project slightly, contain two common rooms next the fresh-air galleries. On the ground floor in the centre are the rooms for the chief medical officer and the matron. Corresponding parts in the two next floors (fig. 30) are occupied by two large dining saloons. There are also a nurse's room and a small ward kitchen on each side and on each floor. The rest of the building on the south side is occupied with bedrooms. These open into well-lighted corridors, two metres wide, which have no other rooms to the north, excepting at the ends, which contain lavatories and closets, and in the basement also a mortuary.

The building was originally intended for seventy-five, the rooms being arranged for one, three, and five beds respectively. Owing to the pressure on their space, however, extra beds have been put in to the number of eighty-eight, so that there are now ten single-bedded rooms, the rest each containing four or six beds. Both sexes are admitted, the men being housed in the western half, the women in the eastern half, with separate sun galleries and dining rooms. There are large verandahs in front of the latter.

The building is heated by low-pressure steam. For ventilation, the windows are constantly left open, day and night. There are ingenious wooden shutters attached to the windows, worked from inside, which admit air while they exclude rain, and can also be arranged as sun blinds.

Only early cases are supposed to be admitted, but the medical officers, in their report, complain that many unsuitable cases are sent in. *Sputa* are received into Dettweiler's flasks, which are provided free of cost. The use of handkerchiefs for this purpose is forbidden. Sputa are afterwards mixed with peat powder and burnt. Linen is

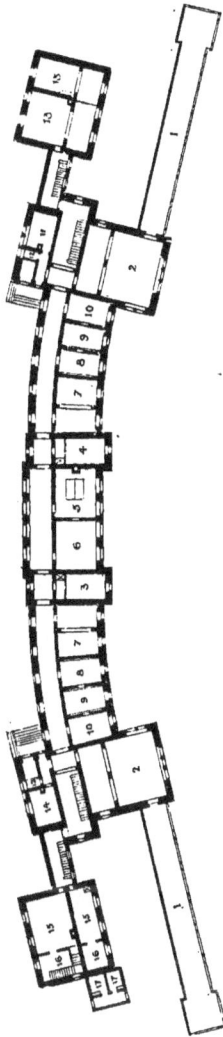

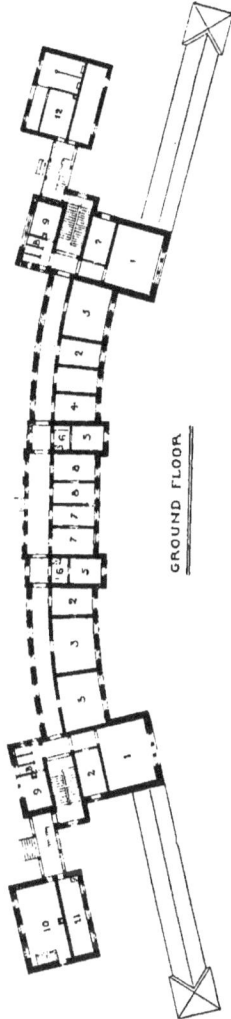

Fig. 29.—RUPPERTSHAIN SANATORIUM.

BASEMENT

1. Fresh-air Galleries.
2. Sitting Rooms.
3. Cellar.
4. Heating Apparatus.
5. Boiler Room.
6. Coal Cellar.
7. Cellar.
8. Dressing Room.
9. Douche Room.
10. Doctor's Room.
11. Earth for Closets.
12. Earth Closets.
13. Laundry.
14. Mortuary.
15. Scullery.
16. Small Kitchen.
17. Pigstye.

GROUND FLOOR

1. Large Dormitories.
2. Single-bedded Rooms.
3. Larger Bedrooms.
4. Drug Room.
5. Nurses' Room.
6. Small Kitchen.
7. Matron's Rooms.
8. Medical Officer's Rooms.
9. Bathroom and Lavatory.
10. Kitchen.
11. Servants' Hall.
12. Coachman, next stables.
B = Bathrooms.

[Face page 254.

disinfected by steam. Patients have the use of the public woods on the other side of the road. Their treatment is as at Falkenstein in all essentials (see p. 177). Smoking is forbidden.

Patients stay at least twelve weeks, if suitable. They pay 5 marks in the private rooms, 3 in the common rooms, or 2½ when they come from one of the Frankfort sick benefit societies. They have five *meals* per diem. They are expected to make their beds, brush their clothes and boots, etc., or else pay 3 marks per month for attendance. There is a good water supply. Sewage is treated with peat mould; waste water disposed of by irrigation. The daily cost is 2·77 marks, of which 1·22 is for food. The average receipts are 2·50˙ marks, so that a deficiency remains to be made up by voluntary subscriptions, in which the Falkenstein patients share.

Dr. Nahm is the chief medical officer, Dr. Fohrbrodt the second medical officer.

In 1895-6, 249 men and 64 women were received. Of these, 179 men and 26 women came through various societies. The percentage of improvement amongst those who left the institution was 77·6.

The nearest station is Eppstein on the line from Frankfort to Limburg.

CHAPTER XL.

SEVERAL large sanatoria are being erected in this part of
Germany, some of which are nearly completed. Nuremberg
is building a sanatorium in the Engelthal, near Hersbrück,
for sixty patients; it is so planned as to be capable of extension to 100 beds, and is estimated to cost 276,000 to 280,000
marks.[1] Stuttgart has a convalescent home (*Neustädtle*)
which receives consumptives amongst other patients, but is
planning a separate sanatorium for 100 beds.[2] There is an
open colony for consumptives at Reichelsheim in the Odenwald, utilised by the Hesse Darmstadt Insurance Company.
A sanatorium is projected at Mühlhain in Rhenish Hesse
Darmstadt.[3] According to Kuthy[4] a sanatorium is projected at Felsberg in the Odenwald for the city of Worms.
Another sanatorium, for the city of Würzburg and Lower
Franconia, is being built at Lohr in the Spessart. In the
Palatinate there is a small sanatorium at Dannenfels for
a large manufacturing firm of Ludwigshafen; and another
for Speyer is being erected near Abbersweiler.[5] In Alsace
and Lorraine there is a project for the erection of a sanatorium for the city of Metz, rendered possible by a legacy
of 150,000 marks from a lady.[6] Saarbrücken is negotiating
for the purchase of a site of 10 hectares (24 acres) in the
Stiftswald at St. Arnual.[7] In the Black Forest Dr.
Baudach has for some years had a sanatorium of this kind

[1] *Das Rothe Kreuz*, 1897, No. 12. [2] Liebe, *loc. cit.*
[3] *Heilst. Corr.*, July, 1898. [4] *Loc. cit.* [5] *Loc. cit.*
[6] *Das Rothe Kreuz*, Sept., 1898. [7] *Ibid.*

under his care at Schömberg, in addition to his private sanatorium. Another sanatorium was opened last year at Arlen
by a local manufacturer for the benefit of his workmen.
A large sanatorium is being built at Marzell on the Blauen
for the Baden Insurance Company. Another is projected
for "semi-necessitous" patients at Freudenstadt in Wurtemberg. Four gratuitous beds are said to exist at the Nordrach Sanatorium (see p. 190). In Bavaria a large sanatorium has just been completed at Planegg Krailling, and
another of still larger proportions is being built for the
Munich Town Council at Harlaching. There is also a
project for the establishment of a sanatorium for Lower
Bavaria at Maxhofen near Deggendorf,[1] and another is to
be erected in the Bavarian forest for the poorer middle
classes.[2]

THE DANNENFELS SANATORIUM.

The *Badische Anilin und Soda Fabrik* at Ludwigshafen,
on the Rhine, are remarkable for the care which they take
of their large staff of workpeople. They have over 4000
workpeople, whom they have provided with a number of
model dwellings, a large restaurant where they can obtain
meals below cost price, another dining hall where those
who receive their meals from home may eat them in company of their wives and children; baths, lavatories and
dressing rooms, with soap and towels gratis; a large free
bath for the wives and children (who are not employed in
the factory) ; a lying-in cottage ; a school of cookery and
household management for the daughters; a cottage hospital for women and children, and another for men; a
convalescent home in the country, and a sanatorium for
consumptives. In addition to this they have a system of
premiums for long service from five years upwards ; a
savings bank ; a sick fund, also open to those who are
called out on military service; a fund for widows, orphans

[1] *Das Rothe Kreuz*, Sept., 1898.   [2] *Heilst. Corr.*, March, 1898.

17

and invalids; and a pension fund for those over sixty years of age.

The convalescent home and the sanatorium for consumptives are bóth near Kirchheimbolanden in the Pfalz. We need only describe the latter.

It is situated at Dannenfels, at the foot of the Donnerberg, 400 metres (1312 ft.) above the sea-level, on 25 hectares (62 acres) of land, which is partly meadow and garden, partly covered with chestnut woods. It is on the eastern side of a hill, and is protected by woods to the north. The eastern end is a three-storey pavilion (fig. 31), with a ground floor, first and second floors and attics. From this extends a long two-storey prolongation, consisting of a single row of rooms on each floor, with a corridor along the northern side, and surrounded on the east, south and west by a balcony with two fresh-air resting places on a level with the first floor. One of these resting places (which has no roof) extends southwards over a built-out pavilion near the eastern end, 10 metres long by 6 wide (32' 9" × 19' 8"). Under it is another shelter open to the west, and with windows to south and east, for use in cold weather. The other first floor shelter is somewhat larger, and occupies the western end of the building, on a level with the garden. The ground floor contains, in addition to the cold weather pavilion, a large dining room at the eastern end, a day room, bath room, doctor's room, ironing room and cellars; and on the northern side, the kitchen and administrative rooms at either end of the house. The first floor contains patients' rooms along the south side; office, staircase, lavatories, closets, and nurses' rooms on the northern side. In the second floor of the eastern end are two rooms for the matron and a reserve room; the attics being occupied by additional administrative rooms. There is altogether room for eighteen patients, six in one room, three each in three other rooms, and three in single-bedded rooms. There is also an outbuilding for cows, pigs, hens and laundry.

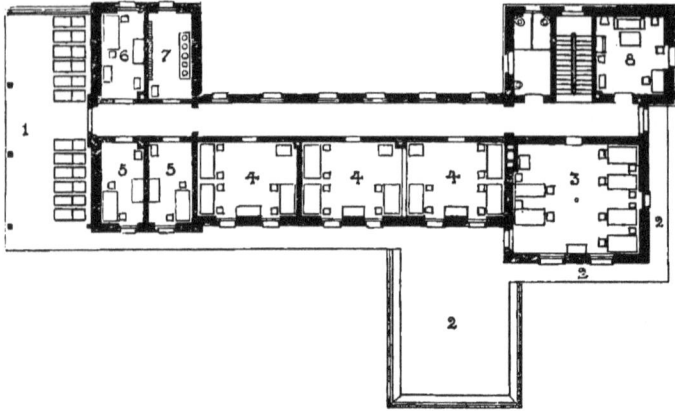

FIRST FLOOR

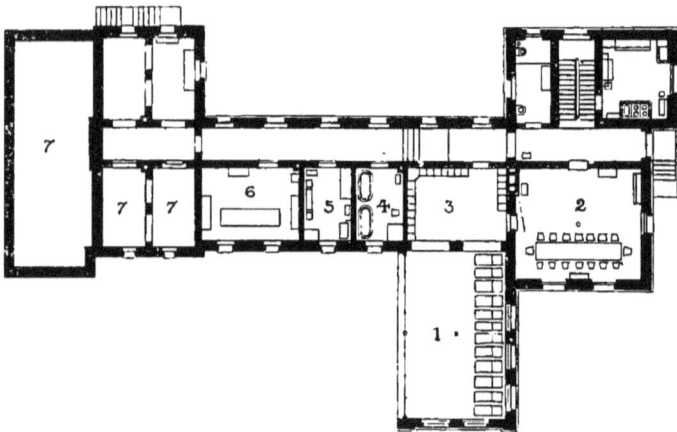

GROUND FLOOR

FIG. 31.—THE DANNENFELS SANATORIUM.   [Face page 258.

Ground Floor :—

1. Fresh-air Gallery.	5. Consulting Room.
2. Dining Saloon.	6. Ironing Room.
3. Sitting Room.	7. Cellars.
4. Bathroom.	N. E. corner, Kitchen.

First Floor :—

1. Fresh-air Gallery.	5. Single-bedded Rooms.
2. Open Platform.	6. Nurses' Room.
3. Room for six Beds.	7. Lavatory.
4. Rooms for three Beds apiece.	8. Office.

The sanatorium was opened in September, 1892, and remains open the whole year. It is under the care of a sister of the *Bavarian Frauenverein vom Rothen Kreuz*, and of Dr. Boyé of Kirchheimbolanden. Up to the end of 1895 fifty patients had been received, of whom ten were still in the establishment. They had received 7230 days of treatment, making an average of $144\frac{1}{2}$ days apiece. Patients now usually stay six months; and the results are stated to be notably better than with only thirteen weeks. Of forty who left, eight remain perfectly well, fifteen much better and able to work; sixteen were no better, and one died. About $\frac{1}{2}$ per cent. of the workpeople become consumptive. The building cost £7000 to erect; the daily cost is 4·15 marks per head, food alone 2·20 marks, doctor and drugs 55 pf.

Patients are treated on the usual lines. They do some of the light work when fit for it.

THE WÜRZBURG SANATORIUM

has been founded by the *Verein zur Gründung von Curanstalten für unbemittelte Lungenkranke* for Lower Franconia and Aschaffenburg, in the Lichterwald, near Lohr, in the Spessart, a district famous for its extensive forests of oak and beech trees, and will have thirty beds. Kuthy gives a description and plans in his book,[1] from which it appears that the sanatorium will have a basement, ground floor, first and second floors. The basement will contain various cellars and store rooms, together with kitchen and laundry. On the ground floor will be a dining room for the men at one end and one for the women at the other, with laboratory, consulting and drug room between, and bath rooms and rooms for servants behind. The fresh-air verandahs will diverge from the ends next the dining rooms, each forming an obtuse angle with the front of the building. On the upper floors there will be nurses' and

[1] *Loc. cit.*

linen rooms behind, and bedrooms in front, some single-bedded, some for three, others for four beds each.

The site is said[1] to have cost 20,000 marks, the estimated cost of the whole sanatorium amounting to 225,000 marks, of which the Insurance Company for Lower Franconia advance 70,000 marks for ten years at 2 per cent. There are at present twenty-two branch societies in the Sanatorium Society, with an income of nearly 120,000 marks.

THE SPEYER SANATORIUM

is shortly to be built on a wooded hill, 380 metres (1245 ft.) above the sea-level, near Albersweiler in the Palatinate. The society (*Verein für Volksheilstätten in der Pfalz*) has about 100,000 marks in hand, besides which the Insurance Society of the Palatinate will lend two-thirds of the estimated cost at low interest.[2]

THE SCHÖMBERG PEOPLE'S SANATORIUM

was originally founded by Dr. Baudach, the owner of the private sanatorium described at page 196. It consists of the old sanatorium, together with a number of other houses in the village with which arrangements have been made. It now accommodates from sixty to sixty-five patients, mostly from the Wurtemberg and Baden, Alsace-Lorraine, Pfalz, and Hesse Darmstadt Sick Assurance Societies. There is an experienced married manager in charge. The patients are kept quite separate from those of the private sanatorium both in the woods and fresh-air galleries, but are otherwise treated in much the same way. Never more than three sleep in one room. The charges are 3 to 3½ marks per diem, or 4 in single-bedded rooms: there are also a few gratuitous beds. In 1896, 297 patients were received.[3] For other particulars, see description of the private sanatorium (p. 196).

---

[1] *Das Rothe Kreuz*, 1st Oct., 1898.
[2] *Heilst. Corr.*, 1st July, 1898.   [3] Liebe, *loc. cit.*

The Arlen Sanatorium.

In the south-west corner of the Black Forest, in the Grand
Duchy of Baden, to the south of Hohentwiel, is a small
sanatorium for male consumptives in connection with the
Heinrichs Hospital, both of which were founded by the
factory owner, Commercienrath Carl Ten Brink, for the
benefit of his workmen, the hospital in 1888, and the sanatorium in 1897. Both are under the care of Dr. Weibel, and
are chiefly intended for the cotton spinning and weaving
operatives of Arlen, although others can also be received.
The sanatorium stands about 400 metres (1300 ft.) above
the sea-level, in open country, at the foot of a mountain
running from north-west to south-east, which shelters it from
north and east winds. It has a fine garden of 1·2 hectares,
and is surrounded by other gardens. It consists of one storey
on a high ground floor, arranged in two symmetrical halves,
for eight male and eight female patients. On each side are
a large day room, one dormitory for four beds, two rooms
with two beds apiece, and a lavatory. The bedrooms have
an average space of 48 cubic metres (nearly 1700 cub. ft.)
per bed. There are a common dining room, kitchen, matron's
and doctor's rooms, and two bath rooms. On the south side
of each half is a large verandah, and smaller ones to east
and west, besides a long covered walk in the garden. The
building is heated by warm-water pipes and lighted by
electricity. Every room has a chimney for ventilation.
The walls are enamel-painted. The sewage is carried into
a neighbouring stream. *Sputa* are poured into the water-
closets without disinfection. Linen is purified by steam
disinfection.

Although mainly intended for consumptives, it is not
reserved exclusively for such patients. The douche is
not used, but other forms of hydrotherapy are employed.
Specific and stimulating remedies are administered when
advisable, but no cod-liver oil. Five *meals* per day are
provided; alcohol being given when specially ordered by

the doctor. There are two female and one male nurse.
Patients pay 2½ marks per day.
The nearest station is Arlen-Rielasingen, on the line from
Singen to Winterthur. Singen is a junction on the line
from Cassel, and that from Basel to Constanz.

THE MARZELL SANATORIUM.

A sanatorium is projected for the *Baden Invaliditäts
Versicherungs Anstalt* at Marzell on the Blauen, 840
metres (2750 ft.) above the sea-level, at an estimated cost
of £33,000, exclusive of electric lighting, and of £4200 for
the furnishing. Of this amount the State gives £2500, and
an annual subvention of £500.

The projected building will be 87 metres long and 10
metres deep. Bath and inhalation rooms, ironing room
and laundry, and the heating apparatus will be placed on the
ground floor. There will be 108 beds, for men only, on three
floors, eighteen rooms containing four each, sixteen being
double-bedded, and four single-bedded ; the cubic space for
each of these sets of rooms being 27·60, 28, and 56 cubic
metres (975, 989, and 1978 cub. ft.) respectively. On the
north side will be lavatories, closets and a day room ; and in
the outbuilding, cellars with the porter's rooms. On the first
floor will be rooms for the medical officers, laboratory, the
office and the kitchen. On the second floor, rooms for the
matron and linen maid, linen rooms, dining saloon and bedrooms for the staff. On the third floor, chapel, library and
guest rooms for the officers of the insurance society. In
front of the ground floor will be a deep verandah for 100
places. The building is to be solidly built, and provided
with central heating and electric light. The soil pipes will
end in iron receptacles accessible on all sides. The floors in
the rooms will be covered with linoleum, in the corridors
paved with " Mettlacher Platten ". There are also to be a
doctor's house, laundry, stables and an ice cellar.[1]

[1] Liebe, *loc. cit.*

Fig. 31*.—The Krailling Sanatorium.

[Face page 263.

The Baden Sickness Insurance Company in 1894 had 1398 cases, of which 546, or 39 per cent., were phthisical or threatened with consumption.[1]

THE KRAILLING SANATORIUM,

which has just been built by the Sanatorium Society of Upper Bavaria to the west of Planegg, is intended for the reception of consumptives of the male sex. Besides working men belonging to the sickness insurance societies (especially that of Upper Bavaria), it will receive private patients of the industrial and commercial classes—shopkeepers and the like—at low fees ; and a number of separate rooms with one or two beds each have been provided with this object. The Sanatorium Society itself, which is under the protectorate of H.R.H. Prince Ludwig of Bavaria, was mainly established through the efforts of Prof. v. Ziemssen in 1895. Amongst the most prominent supporters has been the Baroness Hirsch, who gave a donation of £5000.

The sanatorium stands in a wooded neighbourhood on gravel soil 557 metres above the sea-level, and has an area of 16½ hectares (40 acres) belonging to it, laid out in a variety of paths. It consists of a main building with diverging wings, connected by a subterranean passage about 200 feet long with a subsidiary block. The main building has a basement, ground floor, first and second floors. In the basement on the north side are the kitchen, scullery, and larder. Next to the latter, in the angle of junction between the centre and the western wing, is the ice cellar. Electric ventilation shafts go from the kitchen to the highest point of the building, and lifts to the serving room and to the tea kitchens on the upper floors. To prevent the entrance of cellar air into the upper floors there are no cellars under the parts occupied by the patients, and the cellar stairs do not lead into the shaft of the main staircase but into the open air. In the middle of the ground floor on the south side is the large dining saloon, with a serving room

[1] Kuthy, *loc. cit.*

and servants' dining room to one side and a reading room to the other. The dining saloon is also used as a day room. In the wings are two fresh-air galleries, with room for forty couches. Behind the western wing is a corridor with a number of rooms for the sisters ; and in the angle above the ice house is the Catholic chapel. The corresponding angle on the eastern side is occupied by a Protestant chapel. Behind the eastern wing is a corridor with laboratory, consulting and waiting rooms, and separate bath rooms for the patients, the medical staff and the sisters, as well as a douche room. Part of the corridor is used as a dressing room. To the north of the centre block, behind the dining saloon, is the main entrance, with a visitors' room on one side and two rooms for the house master on the other. Outside these are water-closets, and outside these again, rooms for clean and dirty linen, separated from the two chapels by the main staircases.

The first floor contains patients' rooms on the south side, and at the back of the centre block a committee room, lavatory, tea kitchen, and another large room, besides the water-closets and linen rooms, as on the floor below. The second floor is somewhat similarly arranged. The patients' rooms have an average air space of 38 cubic metres (1340 cub. ft.). Those in the second floor in the wings have mansard windows. As originally planned, there were two rooms with five beds, fourteen with four beds, four with three beds, fourteen with two beds, and eight with one bed apiece ; but 120 can actually be accommodated in the building. The linen rooms are so arranged that dirty linen can be sent down from each floor on one side of the building by a lift into the basement, and clean linen be drawn up by another lift on the opposite side. The building is heated by low-pressure steam, and lighted by electricity. The outbuilding contains in the basement the disinfector with two approaches and bath, another bath room for servants, heating apparatus and cellars. On the ground floor are the washhouse, stables, cowsheds, workshop, tool-house, laundry, engine and dynamo for electric light, mortuary, and common room for men servants. Above are the servants' bedrooms, in addition to hay loft and dry room, the latter with lifts to the disinfecting room and the laundry.

The treatment is on the usual lines. The douche is regularly employed, cod-liver oil and specifics when required. *Sputa* are destroyed by burning.

There are ten nursing sisters. Dr. G. Krebs is the medical officer. Patients pay 3½ marks per day.

The sanatorium is twenty minutes' walk from the Planegg station on the line from Starnberg to Munich, which is distant two hours by rail.

THE HARLACHING SANATORIUM

is being built by the Munich Town Council for convalescents of both sexes, including consumptives, and will probably be opened early in 1899. Situated in 10 hectares (24 acres) of ground, it is surrounded by woods to the south, west and east.

It is to be a large two-storey building, 350 feet long, and nearly 130 wide, with a southerly aspect, and separate administrative block to the north. The latter has been arranged to provide for 500 patients, but the sanatorium will at first accommodate from 200 to 250. The sexes are separated by a division through the building, garden and park, with separate open fresh-air galleries and resting places. The entrance is behind; in front will be a large lawn, and vegetable gardens to the sides of the grounds. Next to the entrance are rooms for the porter and the physician on duty. There will be a separate staircase on each side, as well as one in each wing; also a couple of lifts for patients and others for food. On either side of the main entrance will be a dormitory with twenty beds. In the first and second floors will be other dormitories for twelve beds apiece, besides single-bedded rooms and large open covered balconies well protected from wind by the projecting wings. A chapel and a saloon for Protestant services will be placed on the first floor, the quarters for the nursing staff on the second floor. The floors will be *betonirt*, and covered with linoleum; numerous ventilating shafts will be provided, ending in the chimneys above the roof. The place will be heated by low-pressure steam pipes, lighted by electricity.

The administrative block will contain a very large kitchen, as well as a scullery, dining room for the sisters, doctor's room, laboratory, drug room, and porter's bedroom. The upper floor will contain twenty rooms for the house master, senior physician, and three assistant physicians, and in a separate portion for the sisters. The food will be drawn by mechanical means from the kitchen, through a subterranean passage to the main block. The water supply will be that of Munich. In the wings will be bedrooms for servants, etc. There will be stalls for sixteen cows. In a separate building there will be a laundry with steam disinfector and quarters for the engineer and his mate.[1]

[1] *Das Rothe Kreuz*, 1898, No. 16; Liebe, *loc. cit.*

CHAPTER XLI.

NORWAY AND SWEDEN.

THE movement for the establishment of sanatoria is also in full swing in Norway and Sweden. There have been two sanatoria for paying patients for some time past at Tonsaassen and Gausdal, and a sanatorium for the poorer classes has been established at Reknaes, near Molde, in one of the disused leprosy hospitals, and others are planned at Bergen and in the mountains. The popular present to the king, to commemorate the twenty-fifth anniversary of his accession, amounting to upwards of £100,000, has been chiefly devoted by him to the establishment of such sanatoria. Dr. Armauer Hansen has been largely instrumental in forwarding the movement.

There are also projects for the establishment of sanatoria in Sweden (Kuthy, *loc. cit.*).

NORWEGIAN SANATORIA.

	Feet.	Beds.
Tonsaassen . .	2050	90 (35 to 40 in winter).
Gausdal . . .	2625	?
Reknaes . . .	984	90-100.
Bergen . . .	?	?

TONSAASSEN SANATORIUM

is on the road between Laerdalsören on the Sogne Fjord and Christiania, not far from Odnaes on the Randsfjord. Originally a summer bathing resort for many ailments, it has been of late years more and more frequented by consumptives, who have recently formed as much as 60 per

cent. of the visitors, so that it has been somewhat altered
in character to suit their requirements. The climate is
very good for this purpose, being cool and sunny in summer,
dry and cold in winter. It is less pleasant in April when
the snow melts, and in autumn which is foggy; but the
true winter and the months of February, March and June are
especially pleasant. The temperature ranges in winter
between 28° and 12° F., with a daily variation of 3° to 11°,
and occasional drops below zero. The summer temperature
ranges from 46·4° to 53·6°, with a very regular daily variation of 3° to 5°. The usual maximum is about 68° in June
or July; the lowest recorded temperature was—16·6°, the
highest 80°. The relative moisture is from 85 to 95 per
cent. in winter, 70 to 85 per cent. in summer; with more
in March and April. There are about 135 rainy or snowy
days, and thirty to forty cloudy or foggy days per annum.
There is very little wind, especially in winter. The climate,
as a whole, has been compared by Dr. Andvord to that of
Wildbad; April, May and June he likens to the winter of
Meran; the rest of summer and the autumn to the winter
of Pau; the winter to that of Davos. There are six months
of continuous snow. The sanatorium lies in a hollow in
Tonsaassen Mountain, which separates Valders from Etnedal.
The mountain rises to a height of 2625 feet, the sanatorium
being 625 metres (2050ft.) above the sea. The ground forms
part of a plateau which is surrounded to the west, north and
east by heights of 4900 to 6500 feet, and gradually drops to
the south and east towards the great lakes and sea coast.
The soil belongs to the Cambrian system, and consists of
slate and blue quartz. The latter is somewhat barren, and,
being much fissured, dries rapidly after heavy rain.

The sanatorium owns 3250 acres of land, mostly pine woods
provided with paths at various gradients and many seats.
It consists of a curhaus, six other buildings, doctor's house,
bath house, bakery, outhouses, etc. (fig. 32), mostly arranged
round the head of the little valley. The chief building has
an external staircase, and contains a dining room, music

FIG. 32.—TONSAASEN SANATORIUM.

[Face page 268.

room, conversation room, etc., on the ground floor : it is heated by open fires burning wood. The dining room is 11 × 10 metres (36 × 32 ft. 9 in.) and 3½ metres (11½ ft.) high ; the drawing room of the same size, but 6 metres (19¾ ft.) high. There is accommodation for ninety patients in summer, thirty-five to forty in winter. Very few of the summer rooms have stoves : the others have closed stoves. The bedrooms are in four different buildings, so that four tariffs can be observed. In one of the buildings no tuberculous patients are admitted ; in the other buildings some rooms are used exclusively for tuberculous, others exclusively for non-tuberculous cases. The rooms in one building are 4·5 × 4·1 metres and 3·3 high (14¾ × 13½ × 11 ft.). The rest are about 4·4 × 3·5 × 3 metres high (14½ × 11 × 10 ft.). There are verandahs or balconies on every floor. The furniture is simple. The lighting is by electricity. Ventilation by open windows, day and night, summer and winter. There are said to be good water-closets and baths. The waste water is carried into a brook. In winter the sewage is covered with earth.[1]

The establishment is open throughout the year. It was built in 1881, and has been a winter station since 1885. No advanced cases are admitted. Treatment is by open air, in the verandahs or balconies, or in the pavilions in the woods. Patients who are fit for it take plenty of exercise. There is a very complete apparatus for hydrotherapy, with vapour baths, needle baths, ferruginous, hot and cold baths, etc. Patients in summer have friction with water at 15° to 20° C., or douches. In winter, dry friction and partial ablutions are substituted. Five or six *meals* are provided daily, with alcohol in great moderation. Cod-liver oil and specifics are little used. The *sputa* are put into a cask with sol. of ferrous sulphate, and after a month are burnt. Patients bring their own bedcovers and pillows. Mattresses are disinfected by brushing with 2 per mille corr. subl., followed by solution of washing soda. Rooms are

[1] Möller, *Les Sanatoria.*

rubbed with bread and then washed with soap and water. There is one nurse.

Patients pay in summer 120 to 175 kroner[1] for four weeks ; in winter, 150 to 180 kr. This includes room, pension, medical attendance, baths, light and heating. Beverages are extra. Casual visitors pay a little more. There are a post office and telephone at the sanatorium, which can be reached in a day from Christiania by rail, steamer, and cariole, *viâ* Randsfjord, Spirillen, or Mjösen. The route from Bergen is *viâ* Laerdalsören.

Dr. Jac. Sömme has been the medical officer for the last two years, in succession to Dr. Andvord.

THE GAUSDAL SANATORIUM.

According to Daremberg,[2] there is another sanatorium for consumptives at Gausdal, near Lake Mjösen, 800 metres (2625 ft.) above the sea-level.

THE REKNAES CONSUMPTION HOSPITAL,

near Molde, on the north side of the beautiful Molde Fjord, was converted by royal decree in 1897 into a consumption hospital for poor people. Molde is sheltered by mountains to the north and west, and has consequently a milder climate and more luxurious vegetation than most other parts of Norway. Roses and honeysuckle bloom freely, and horse chestnuts, lime trees, ashes, and cherry trees are found as well as the usual birches and pines. Stor Tuen, behind Molde to the north-east, has a height of 3200 feet ; while to the east and south are mountains of 5000 and 6000 feet, with the magnificent scenery of the Romsdals Fjord.

The hospital is situated in a large garden on a knoll overlooking the fjord, a little distance westwards from Molde, at an elevation of 300 metres (984 ft.) above the sea. It consists of a wooden building on two floors, composed of a centre and two projecting wings, which together enclose a courtyard, and are separated by a road in front

---

[1] A krone is worth about 1s. 1d. ; 150 kr. = £8 6s. 8d. ; 180 kr. = £10.

[2] *Les Etablissements fermés pour le Traitement de la Tuberculose*, Paris, 1892.

from a long open shelter or *liegehalle*. It has accommodation for ninety to 100 patients, with common rooms of about 200 cubic metres (7060 ft.) capacity, bath and douche rooms, and a number of bedrooms, each for two or four patients. These have all of them chimneys, and have an average air space of 140 to 150 cubic metres (4950 to 5300 cub. ft.). Ventilation is chiefly by open windows. Sewerage is carried into the sea.

None but consumptives are admitted, the medical director having power to decide which cases have the best claim for treatment, which lasts as a rule for at least three months. Random spitting and smoking are strictly forbidden. *Sputa* are received exclusively into Dettweiler's flasks or spittoons; they are disinfected by boiling and then thrown into the water-closet. Linen is disinfected by heat in a disinfector. The rooms are disinfected by washing. The douche is occasionally used; sometimes cod-liver oil, but no specifics. Five *meals* a day are provided, alcohol being given where necessary.

There are four nurses and ten servants. Dr. Edv. Kaurin is the sole medical resident, and has supreme command.

The charges, payable by assurance societies, private friends, or others, are 1·20 kr. per diem. The remaining expenses are borne by the State and the municipality.

The place may be visited from England by steamer from Bergen, Christiania or Trondhjem.

OTHER SANATORIA FOR THE POOR

are being established in various parts of Norway. One of these for eighty beds is to be started at Bergen in the disused St. Jurgen's Leprosy Hospital; and another for eighty to 100 patients will be erected somewhere in the mountains. The situation at Bergen is not considered a very good one, as the building dates from the sixteenth century, and is in the midst of a town of 50,000 inhabitants, while the climate is exceedingly rainy though mild and equable. The annual rainfall is over 72 in. Dr. Holmhoe is to be chief medical officer (Kuthy).

CHAPTER XLII.

Up to 1892 there were two sanatoria in Russia for consumptive paying patients, one at Halila in Finland and another at Lindheim in Livonia. The first of these however was transformed by the late Emperor Alexander III. into a popular sanatorium, and two others built in the same grounds respectively for girls of the upper classes who show consumptive tendencies and for military men. There are at present, in addition to the above, sanatoria for paying patients at Slawuta in Volhynia : at Yalta in the Crimea ; at Oranienbaum near Peterhoff, opposite Cronstadt, for patients with but moderate means : and a number of koumiss stations on the Volga and near the Steppes ; while a new sanatorium for paying patients is to be erected under Dr. Masing at Willmanstrand on the Saimansee in Finland.[1] The sanatorium of Slawuta is near Ostrog on the Goryn, and has 100 beds, but is only open in summer. For the poorer classes a sanatorium has been founded at Taitzi on the Baltic by the present Emperor : consumptive patients are also received at the Obouchowsky Hospital at St. Petersburg (for 100 men and fifty women), the Alexandre Hospital in the same city (for fifty men and fifty women), the Alexandrina Hospital (for fifty women) ; and " House Sanatoria " have been established for the fresh-air treatment of consumptives at the Military Hospital of Zarskoje Selo under Dr. Unterberger, at the Hospital of the Grand Duke Nicholas at Kief, and at Wola near Warsaw under Dr. Natanson : while other sanatoria are to be erected near

[1] See Liebe, *Hyg. Rundsch.*, 1896, No. 20.

(272)

Moscow and at Mustamakki in Finland.[1] The Duodecim
Society of medical men in Finland has recently decided to
erect a sanatorium for fifty to sixty patients. It is estimated to cost 50,000 to 75,000 rb., to be covered by shares
of 100 rb. each.[2] The Rachmanow family in Moscow has
made a donation to the city of 200,000 rb., part of which is
to be for the erection of a sanatorium of at least 100 beds,
while the interest on the rest is to be spent in the support
of free beds in the sanatorium. The city of Moscow is
expected to pay for the maintenance of the institution, and
to provide a suitable site.[3]

A society has been formed in St. Petersburg for the erection of sanatoria for tuberculous children; 22,000 rb. have
been collected.[4] Countess Bariatinsky has recently founded
a gymnasium or school in Yalta for sixty children who have
to live in the south owing to delicate health.[5]

Collections are to be made twice a year in all the churches
of St. Petersburg for the erection of sanatoria. The first
collection yielded 1200 rb.

RUSSIAN SANATORIA.

		Feet.	Beds.
Lindheim	Livonia	300	12
Quisisana, Yalta . . .	Crimea	200	? 36
Alexander, Halila . . .	Finland	—	32
Maria ,, . . .	,,	—	25
Nikolaj ,, . . .	,,	—	100
(Willmanstrand) . . .	,,	—	—
Slawuta	Volhynia	—	100
Oranienbaum	Baltic	—	—
Taitzi	,,	—	40
(Moscow)		—	100
Zarskoje Selo (House San.) .		—	—
Kief ,, .		—	—
Wola (Warsaw) ,,		—	—
Obouchowsky Hospital . .	St. Petersburg	—	150
Alexandre ,, . .	,,	—	100
Alexandrina ,, . .	,,	—	50

[1] Kuthy, loc. cit. ; see also Léon Petit, loc. cit. ; Sonderegger, Heilstätten
für Brustkranke in der Schweiz, St. Gall, 1894.

[2] Heilst. Corr., Aug., 1898. A rouble is worth about 2s.

[3] Ibid.   [4] Ibid.   [5] Ibid.

18

LINDHEIM SANATORIUM,

in Livonia, was originally a boarding school, but was acquired by Dr. Armin Treu in 1895, and opened as a sanatorium for middle class consumptives. It consists of two wooden buildings, with one and two floors respectively, standing in a garden of 5 hectares (12 acres), nearly 100 years old, and wooded with oaks and other deciduous trees. There are altogether 75 hectares (185 acres) of land belonging to the sanatorium. The soil is of granite and sand without any admixture of clay, so that it rapidly dries after rain.

The sanatorium stands 300 feet above the sea-level, in a plain traversed by the little river Waidau, and surrounded by pine woods to the west and south, and by mountains to the north and east. The sanatorium is said to be warm, and comfortably though not luxuriously furnished, heated with Russian stoves, lighted with petroleum, and provided with special ventilators. The dining and reading rooms cover an area of 96 square metres (1033 sq. ft.), and are 3 metres (9 ft. 10 in.) high. The bedrooms, which are mostly in the two-storey buildings, are $3\frac{1}{2}$ metres (11 ft. 6 in.) high, and vary from 16 to 48 square metres (172 to 516 sq. ft.). For open-air treatment there is a large verandah open only to the south. There is at present accommodation for twelve patients, but Dr. Treu intends to buy the property and completely transform the establishment, increasing the accommodation to twenty-four.

The sanatorium is intended for consumptives in remediable stages; but healthy attendants or friends are also admitted. The treatment is by hygienic methods. The *food* is said to be abundant, with good plain cooking. Wet and dry rubbing are practised, but the douche is not employed. *Sputa* are received into Dettweiler's flasks. The contents are disinfected with $FeSO_4$, and poured into the water-closets. The sewage goes into a cesspool. Linen is purified by boiling; rooms by thorough washing, rubbing with bread, and application of corrosive sublimate solution. Creosote, Kleb's antiphthisin and Kochs' tuberculin are employed in suitable cases. Cod-liver oil is not given. There is a nurse in the building who has been trained by Dr. Treu.

The charges are 50, 60, and 75 roubles per month. If two persons share one room, the first pays the full price, the second 50 roubles per month. For healthy attendants there is a reduction of 10 roubles. For extra nursing 40 roubles are charged in addition. Other extras are beverages and expensive drugs. Patients bring their own bedding and body linen; also a rubbing glove, a fur covering for winter, with fur cap and felt boots to the knees. They buy their own Dettweiler's flask and thermometer.

The sanatorium may be reached by carriage from Werro, which is connected with St. Petersburg by rail.

Kuthy (loc. cit.) mentions a sanatorium at Lindenhof for thirty patients. This is probably the same as the above described establishment.

QUISISANA.

This is a sanatorium not specially for consumptives, but admitting such patients in early stages. Situated in a suburb of Yalta, in a bay on the Black Sea coast, it is protected to the north, west, and partially to the east by the Yaila Mountains, which rise to a height of 5000 feet. The soil is slate, quartz, crystalline limestone and marble, mixed with clay, diorite, etc. The sanatorium is 200 feet above the sea-level, with a view of the sea to the south. It stands in a terraced garden of one-third acre, next to a wood of 7 acres ; and all the neighbouring houses are also surrounded by gardens. The grounds are on the mountain side, and are laid out with paths of various gradients.

The sanatorium, which was founded in 1886, consists of several buildings, one of which is a wooden structure with seven rooms and four verandahs and balconies, the others being solidly built of stone. Higher up is the main block,

which is built in two storeys and contains twenty-four bedrooms with seven verandahs and balconies, the dining room
and drawing room. Still higher is the doctor's residence,
which contains the bath room, laundry and kitchen department, in addition to ten bedrooms with three verandahs and
balconies. A little way off are the stables, cowsheds, etc.
The verandahs have for the most part glazed sides and open
fronts. The average air space in the bedrooms is 1104 cub.
ft. ; the common rooms collectively contain 7000 cub. ft.
There are no ventilating chimneys, but double ventilation
for winter and summer, admitting fresh air through open
windows or warmed air next the heating apparatus. This
consists of Warsaw porcelain stoves with hermetical seal
and chimney. Throughout the winter every window is
opened while the patients are at meals. The lighting is by
stearine candles and petroleum. There is room for thirty
patients. As a rule only one sleeps in each room, but
exceptionally two or even three may sleep together.

The *sputa* are disinfected with carbolic solution and then
poured down the water-closet. Linen of healthy patients is
washed at home ; of those who are ill by the public laundries,
which also disinfect by steam. Rooms are disinfected on
the departure of each patient by the public health authorities, and repapered and repainted of a different colour,
mainly to give confidence to the visitors, the floors being
washed and disinfected with sublimate solution and then
treated with beeswax and turpentine. The solid sewage
and kitchen refuse are removed every day and carried out
to sea by a special steamer. The liquid sewage is carried
along the seashore in ventilated sewers.

Patients in the third stage are not admitted, nor as a rule
any which require nursing. The treatment is by open air,
good food and the like, but not by the Brehmer methods.
Dr. Weber seeks to make the place a home, and to banish
everything suggestive of a hospital or sanatorium. He is
present with his family at all meals, and treats patients
more as members of the family. He constantly endeavours

to impress upon them the very feeble infective power of
tubercular consumption, and organises concerts, private
theatricals, etc., in which both patients and the members of
his own family take part. A masseuse lives in the house ;
other nurses are obtained from the Red Cross sisters in case
of need.

The charges are 70 to 90 roubles per month, and 10 roubles
for medical attendance.

Yalta may be reached from Sebastopol by steamer and
carriage.

THE SANATORIA AT HALILA.

A sanatorium was founded in 1889 by Dr. Dittmann at
Halila in Finland for the reception of paying phthisical
patients. It was however bought by the Russian Government by command of the Emperor Alexander III., and converted in 1892 into a popular sanatorium for the same
purpose. Two other sanatoria have since been erected on
the same ground. These three sanatoria are called respectively the Alexander, the Maria, and the Nikolaj.
The ground belonging to these sanatoria consists of 1000
hectares (2471 acres) of wood and moorland near the Halila
lake: and has no large towns or factories near it. Walks
have been laid out at various gradients provided with seats
at every 50 metres. There are already a great many such
paths, but the number is being added to every year. There
are altogether some twenty buildings on the ground,
including the three sanatoria, which are under common
management.

THE ALEXANDER SANATORIUM

consists of a pine-wood structure on a granite foundation
in the shape of a T, with ground floor and first floor, and
contains thirty-eight rooms for thirty-two patients. Most
of these rooms are east and west of a wide corridor which
leads on to an open verandah on the south side, or on to
balconies above it. The bedrooms are $4\frac{1}{2}$ metres ($14\frac{3}{4}$ ft.)
high and $8 \times 5\frac{1}{2}$ metres (26×18 ft.); and have chimney

shafts for ventilation. Their walls are of varnished pine-wood, the floors covered with linoleum. The furniture is of white enamelled pinewood. There are also common rooms, a music room, a reading room, billiard room, and winter and summer dining rooms, as well as two bath rooms, douche room, consulting room, and a laboratory with chemical, bacteriological, histological and microphotographic apparatus. The kitchen department is united by a warm corridor with the rest of the building. There are balconies to the north as well as south. Heating is by Swedish porcelain stoves; lighting by electricity. There is an abundant water supply, which is mostly pumped up by a steam engine from the Halila lake into two highly placed reservoirs, the drinking supply being stored in a third reservoir. This water supply is common to all three sanatoria, and is capable of supplying from 60,000 to 90,000 litres per diem, or from 400 to 600 per head. The water-closets are of English pattern, with powerful spiral flush, and carry the sewage through earthenware pipes to a small lake adjoining the property, where it is purified by filtration through sphagnum, the solids being put on to the fields.

Patients receive four *meals* a day, in addition to milk or other supplementary food. They spend most of the time out of doors from 8 A.M. till 10 P.M. Morning and evening they are treated with douches, baths and massage. Wet compresses are much used. *Sputa* are received into Dett-weiler's flasks, random spitting being strictly forbidden. Spittoons containing water are also used. The bed and body linen, which with handkerchiefs, etc., is provided by the patient, is disinfected by steam before being washed. There is a good library with newspapers, etc.; patients also amuse themselves with piano, chess, photography, etc.; and in winter there is sledging. They are also encouraged to do light work in the garden; and a bookbinder's apparatus is to be set up for them. Dr Gabrilowitsch is the chief medical officer.

In 1895-6, 109 patients were treated, 43 per cent. being

in the first and 47·7 per cent. in the second stage. Of these 19·3 per cent. were apparently cured, an additional 54·1 per cent. were improved ; in 18·3 per cent. there was no result, and 8·3 per cent. died.

The mean duration of stay was 219 days. The results in winter were on the whole more favourable than in summer. The charges are from 1½ to 2 roubles per day.

THE MARIA SANATORIUM,

which was founded in 1893, is intended for children who have a predisposition to consumption, and especially for the girls of the schools for the nobility who may show such tendencies. The sanatorium contains thirty-five rooms, for twenty-five patients, all placed on the south side of the building, a long, wide, airy, glazed corridor being placed to the north and used in bad weather for walking exercise. The rooms resemble those in the Alexander Sanatorium, but are rather more luxuriously furnished. They have air-shafts for the foul air, and also means of warming the incoming air. The flooring is of parquet. Heating is by Swedish stoves, lighting by electricity, and the sanitary arrangements are like those of the Alexander Sanatorium, but rather more complete. The walls in the kitchen and bath rooms are tiled, the kitchen being placed in the ground floor. The lower floors and the walls have been built with damp-proof courses.

THE NIKOLAJ SANATORIUM

was built in 1895 by the Grand Duchess Alexandra Jossef-owna for the military classes. It is built on the plan of Hohenhonnef (see p. 182), in three storeys, and has 120 rooms for 100 patients. The front has a southerly aspect, and is provided with three tiers of fresh-air galleries and diverging wings. The large dining saloon is over the kitchen in a separate block to the north, united to the main building by a vestibule and winter garden. The walls and ceiling are of varnished wood, the floors covered with linoleum. For ventilation there are separate air-shafts and fans, with electromotor. The heating is by low-pressure warm-water pipes. The foundations have a damp-proof course, the electric lighting, water supply, and sewerage being part of the same system as in the other two sanatoria, and the rooms of the same size. Dirty linen is put into numbered sacks and sent down a linen shoot into the basement to be disinfected and washed.

This establishment is not far from Novajakirka, on the branch railway from Viborg to St. Petersburg.

THE TAITZI SANATORIUM.

Early in 1895 the Society of Medical Practitioners in St. Petersburg made a report on the need for sanatoria for the consumptive poor, and appealed to the public for funds. Plans and estimates were prepared, and ultimately a society was formed under the presidency of Countess Woronzova-Dashkova, and the high protectorate of the Empress Maria Theodorovna. Soon after the Emperor Nicholas II. sent a contribution to the society of 467,000 roubles (nearly £50,000) for the erection and maintenance of a sanatorium in memory of the Empress Maria Alexandrovna, who was herself a victim of consumption. He also made them a gift of his property of Taitzi, near Gatschina, on the Baltic coast, about twenty-five miles from St. Petersburg. This place has long been famous for its magnificent park and its sheltered situation, and contains 22 dessiatinen (630 acres) of arable land, gardens, meadows, and woods. In 1896 extensive alterations were made at the cost of the Department of Imperial Properties, the land being drained, the mill removed, a pond filled in, water-courses cleansed, and their banks turfed. The old stone castle has been fitted for the accommodation of twenty patients, while other buildings have been altered to receive the doctor, nursing staff, and servants. Verandahs have been constructed on every side of the sanatorium, and there is also to be a solarium on the roof. A new building is also being

built for twenty female patients ; and a laboratory, disinfecting chamber, electrical apparatus, laundry, and chapel
are also being added. The china has been sent from the
imperial factory.

MILITARY HOUSE SANATORIA.

These have been chiefly advocated by Dr. S. Unterberger, who was the first to establish one at the Military
Hospital of Zarskoje Selo, near St. Petersburg. Dr.
Unterberger is not a believer in the common causation of
phthisis in all its forms by the tubercle bacillus, but is of
opinion that many scrofulous inflammations of bones, joints,
glands and lungs are non-bacillary, and due to inherited
predisposition : "a constitutional disease, and a specific,
still unknown ailment of the lymphatic system ; just as
hæmophilia and arteriosclerosis are special ailments of the
muscular system". Unterberger believes this tendency to
be greatly encouraged by defective nutrition, noxious micro-
organisms, such as the streptococci, staphylococci, micro-
coccus tetragonus, etc., and that the tubercle bacillus may
secondarily attack such lesions. This view is, it appears
to me, of more theoretical than practical importance, as in
any case treatment would be directed against constitutional
debility, and sputa disinfected by the usual methods.

Dr. Unterberger started his House Sanatorium as follows :
A large airy room was chosen with a southerly aspect, and
a number of rooted young fir trees placed in it in tubs
of wet sand. These were sprinkled daily with pulverised
solution of ol. pini sylvestris, ol. terebinth and water. The
fir trees retain their needles for about six weeks, when
they are replaced. Patients sleep in these rooms, and in
other respects are treated according to the Brehmer-Dett-
weiler methods.

Dr. Unterberger has published statistics showing that
out of 128 patients 8·6 per cent. were restored to compara-
tive health, while 38·3 per cent. more were improved ; a
result which he states is much better than is obtained

by ordinary methods in hospitals with a similar class of patients.[1]

THE WOLA HOSPITAL,

situated in a suburb of Warsaw, has been specially constructed under the direction of Dr. Natanson with the view of providing a specialised out-patient department. It consists of a number of independent blocks scattered over a park of nearly fifteen acres. One of these is intended for consumptives, and is arranged on a southern slope in a horseshoe shape, in such a way that none of the patients' rooms face due north or south. The bedrooms contain at most two beds; they are paved with a special smooth and resisting kind of porcelain, the walls up to a certain height being similarly coated, and all corners rounded. The heating is by means of stoves, which are charged from the corridor outside so as to prevent any dust flying about. The ventilation is said to be very good. The slops and linen are removed through special tubes instead of through the corridors, the linen falling into an antiseptic bath in the basement. Patients stay during the daytime in the gardens, or in the common room, or are wheeled on to the verandah.

The patients are treated gratuitously.[2]

---

[1] See *St. Petersb. Med. Woch.*, 1896, No. 32; *Deutsches Militärärztliche Zeitschr.*, 1898.

[2] Léon Petit, *loc. cit.*, quoted from Jasiewicz.

exist at Arosa, Davos (Dr. Turban's and Dr. Philippi's), Leysin, and Montana. There are also homes for more advanced cases at Davos (Maison de Diaconesses and Villa Pravignan). Many of the hotels at Davos would be entitled to rank with sanatoria but for the absence of medical supervision and the presence of other kinds of visitors and patients. The open-air treatment is also carried out at Les Avants, near Montreux, under the care of some local physicians.

Sanatoria for less wealthy patients are described in the next chapter.

SWISS PRIVATE SANATORIA.

		Feet.	Beds.
Arosa	Cant. de Grisons .	6090	65
Davos, Turban . .	,, .	5160	80
,, Philippi . .	,, .	?	40
Leysin	Cant. de Vaud .	4760	110
Montana	Cant. de Valais .	4970	50

SANATORIA FOR POORER PATIENTS.

		Feet.	Beds.
Heiligenschwendi . .	Cant. de Berne .	3800	52
Davos (Basel) . .	Cant. de Grisons .	5250	70 (100)
,, (Dutch) . .	,,	5120	30
,, English Home .	,,	?	?
St. Moritz ,, .	,,	?	?
Faltigberg (Zurich) .	Cant. de Zurich .	2950	92-100
Stachelberg (Glarus) .	Cant. de Glarus .	3940	30
Aegeri for Children .	Cant. de Zurich .	2690	?

The Arosa Sanatorium.

Arosa is a small hamlet in the Upper Engadine, in the canton of Grisons, about six hours' drive from Chur. The road ascends the valley of the Plessur in an easterly direction as far as Langries, where it takes a sharp turn to the south-west as far as Arosa. The Arosa valley itself forms a long oval, surrounded on all sides by snow mountains, and rises from 1750 to 1892 metres above the sea-level. It is divided by the Tschuggen mountain on the north into an upper and a lower part; and its sides are clothed to a height of about 1900 metres (6200 ft.) by pine woods, broken by two fine lakes.

The climate is Alpine, with a dry diathermanous atmosphere, much sunshine, little or no dust, and a low barometric pressure. Owing to its position among high mountains, it is little affected by strong winds; but is kept fresh and clear by local winds of great regularity. The prevailing day wind in winter is westerly, at other seasons easterly; while at night the cool mountain air flows down into the valley from the north. Excepting when the *Föhn* blows, however, there is no strong wind, so that low temperatures can be comfortably borne. As in all Alpine climates, there are considerable daily and seasonal differences of temperatures; although these are diminished by the abundant vegetation. In winter the minimum and mean temperatures are about 2° C. higher than at Davos; while the amount of sunshine is greater. In summer Arosa is a little cooler than Davos, but has a larger proportion of cloudy days. Altogether the climate of Arosa is regarded by Dr. Ewart as more stimulating than that of Davos.

The sanatorium is situated on the northern side of the Arosa valley, and is sheltered to the north by the Tschuggen, to the east by pine woods and the approximation of the mountains; while it is open to the south and west, and so receives a large proportion of sunshine. It is 1856 metres

FIG. 33.—THE AROSA SANATORIUM.   [*Face page* 285.

(6090 ft.) above the sea-level, and 120 metres above the bed of the Plessur, and was built in 1887, and extended in 1895. It is isolated from other buildings by private grounds, which are provided with a few horizontal paths and a larger number at gentle slopes, all plentifully provided with seats. Some of these walks are over sunny meadows, others through the woods. The view to the south is over meadows which rise to naked peaks in the distance ; to the west are wooded mountain sides. Owing to the steepness and the permeable soil, the ground soon dries after rain and during the melting of the snows.

The sanatorium consists of an old and a new curhaus side by side, and a separate building for laundry and dis- - infection. The ground floor of the old curhaus (fig. 33) is mainly occupied in front by drawing room, music room and reading room, behind by the kitchen department. In front of it is a deep verandah, protected at the eastern end by the projecting music room, and ending at the western end in a pavilion. On the ground floor of the new curhaus, which stands back about 10 metres, are a ladies' room, offices, vestibule, waiting room, consulting room and laboratory ; and on the south front another verandah. Uniting the two buildings is a large dining saloon, which covers 160 square metres (1722 sq. ft.), and is 6 metres (19½ ft.) high. Each building has a terrace in front of it, and at a lower level a *liegehalle*. The united south front measures 105 metres. There are two upper floors in the new kurhaus, and three in the old, which are occupied by bedrooms. Originally opened with thirty-two beds, there are now sixty-five. Most of the bedrooms have at least one window to the south, and are of good size ; some have separate balconies. Every room is lined with wood, and has linoleum-covered floors. The electric light is used throughout the establishment. Heating is by solid closed stoves in the old building, by low-pressure steam in the new. There is a private water supply, and good flush to the water-closets, which are placed in built-out pavilions near the staircases

and bath rooms. The baths and douches are partly in a
built-out pavilion at the western end of the new curhaus,
partly next the kitchen in the older building. Disinfection of linen is by steam. Patients buy their own
woollen bedcover at cost price.

Meals are numerous and abundant. Soup is served before
breakfast ; after this come two breakfasts, midday dinner,
afternoon tea, and early and late supper. About a pint to
a pint and a half of extra milk is supplied to each patient.

The sanatorium is open all the year round, and only
receives consumptive patients. Dr. Jacobi, who is the
resident medical officer, regards the following cases as un-
fitted for treatment at Arosa : those with extensive tubercular infiltration, ulceration of the larynx, chronic anæmia,
pronounced chronic neurasthenia, diabetes, uncompensated
valvular defects of the heart, disease of the kidneys. Erethic
forms of phthisis also do not usually do well ; nor do those
with a rapid course with typical suppurative fever. On
the other hand, hæmoptysis and fever do not contra-indicate,
and Dr. Jacobi does not agree with Dr. Wolff-Immermann
of Reiboldsgrün that improvement begins within the first
fortnight or not at all. In some cases fever goes away for
the first time after many weeks' treatment. Dr. Jacobi
believes in the utility of exercise whenever patients are
sufficiently strong.

The number of patients treated at Arosa during the four
years ending in spring, 1897, was 259, the days of treatment
31,350, or an average of 121 days. Forty-seven patients
(18·1 per cent.) failed to improve, fourteen of them being in
too advanced a stage, or having too feeble reactive powers
to be likely to improve.

The charges for rooms vary from 1·50 to 6 frs. per diem,
or 3 to 7 if double-bedded. Attendance, including baths,
douches, rubbing, and medical attendance, 8 frs. per diem.
An entrance fee of 15 frs. is charged after the first fortnight, and 1 fr. extra per diem in winter for heating, lighting, and attendance.

Arosa may be reached by diligence from Chur, which is connected by rail with Zurich.

DAVOS

is situated amongst the mountains of Grisons in a high funnel-shaped valley traversed by a little torrent, the Land-wasser. The valley runs from north-east to south-west at an altitude of about 1560 metres (5120 ft.) ; it is 12 kilom. long and 1 kilom. wide, and is protected to the north by a high chain of snow mountains, and to the east by a strongly projecting mountain spur. The valley to the south is covered with fields and surmounted by pine woods. At the northern entrance is Davos Dörfli ; then the valley forms an angle and enlarges ; and here is found Davos Platz. Towards the south there is a third village—Davos Frauenkirch ; after which the valley begins to narrow and ends in a narrow wild defile. To the north-east of Davos Dörfli is the Davos lake. Davos Dörfli receives more sun than Davos Platz, but is much less sheltered from the north. On the other hand, Davos Platz has a large population in comparison with its area, many of them consumptives, and is served by a railway from Landquart, so that in winter there is some-times a heavy cloud of smoke over part of the town. The two health resorts are separated by a projecting shoulder of mountain which hides one from the other. A tramway unites them, as well as the railway.

The climate of each is Alpine ; in winter the ground is constantly covered with snow; there is little wind, but much sunshine, and a dry cold atmosphere free from dust ; in summer the place is warmer and more windy. Between the two is the period of the melting of the snows, which used to be the signal for the departure of many patients. Many however now remain until they are well. The place, which was originally brought into notice by Dr Spengler, a Ger-man refugee, in 1848, and afterwards recommended in 1862 by Dr. Unger, one of Dr. Brehmer's patients, has all the advantages of an Alpine climate, together with some of the disadvantages which inevitably exist in a fashionable health resort, where many consumptives congregate together with a mixed population of tourists and invalids. Facilities abound for skating, tobogganing, sledging, snowshoe running and the like ; also for more objectionable concerts and entertainments in crowded places. Many of the consumptives there are scarcely, if at all, under medical control, and are apt to disregard the directions of their medical advisers, to their own hurt and possibly to the disadvantage of others.

There are a number of hotels and pensions, some of them very well arranged, and possessing many of the structural features of medical sanatoria, but without their medical control and attention to details of management. Most of these hotels charge from 7 frs. 50 c. to 14 frs. per day ; in the pensions the terms are from 1 to 2 frs. less. The sanatoria proper consist of (1) Dr. Turban's sanatorium, (2) Dr. Philippi's new sanatorium at Davos Dörfli, (3) a sanatorium for the poor of Basel, (4) one for poor consumptives of Dutch nationality ; we may perhaps add (5) the Diakonissen-Haus, a private nursing home for advanced cases, and (6) the Villa Pravignan, which is a similar institution exclusively for Roman Catholics.

DR. TURBAN'S SANATORIUM

is situated on the south-west of Davos Platz on the mountain side at an altitude of 1573 metres (5160 ft.) above the sea-level. Sheltered by mountains and forest to the north, it has a fine view to the south and south-west over meadows and the town of Davos to the mountains beyond. It has a garden of 7 hectares (17 acres) with walks at gentle gradients, adjoining the grounds of the Kurverein, to which patients have access on payment of a subscription. In these grounds " the paths are somewhat steep, and insufficiently provided with seats, while in summer there is too little shade ".[1]

[1] V. Jaruntowsky, *Sanatoria for Consumptives*, trans. by E. C. Beale London : Rebman Publishing Co.

FIG. 34.—DR. TURBAN'S SANATORIUM, DAVOS, SWITZERLAND.   [*Face page* 289.

The sanatorium, which was built in 1887 and enlarged in 1894, is situated at the lower or south-east corner of the garden, and consists of four buildings (fig. 34) united by covered passages, and an isolated villa for the chief medical officer and the managing director.   The main building is a long erection on four floors, with a basement, having a covered verandah in front of the ground floor, and open balconies to the other floors.   In the basement are the administrative rooms and cellars, the laundry, disinfecting and heating apparatus.   On the ground floor is a large dining saloon, 15 × 10 metres (49 ft. × 32 ft. 9 in.) and 5½ metres (18 ft.) high, with room for 120 guests ; a drawing room, reading room, billiard room, office, doctors' rooms and laboratories, gymnasium, douche and bath rooms.   Above these are patients' bedrooms on three floors, all on the south side, and opening into a well-lighted corridor, the staircase being also to the north.   The largest of the other villas is to the east, the remaining two being at the upper or western end. There is altogether accommodation for eighty patients.   The bedrooms average 57 cubic metres (2013 cub. ft.) in capacity. They are heated in the main building by low-pressure steam pipes, the rest of the establishment being served by porcelain closed stoves or hot-water pipes.   In each bedroom of the main building is a radiator which can be regulated, and which has a ventilating inlet next to it.   On the opposite wall is an outlet leading to a chamber under the roof. In addition to this, the windows, which are French windows with an upper hinged pane, are always kept partly or entirely open.   There is a similar arrangement over the doors of the balconies.   The walls are of wooden panels with rounded angles.   The floors are covered with linoleum. The lighting is by electricity, the bedrooms having bed and table lamps which can be shaded.   The water-closets have a good flush.   There is a very good water supply.

The diet is varied, with six *meals* a day, including much milk and butter.   One day it is *recherché*, the next it is simple in character.

19

The patients are exclusively such as are likely to benefit by an Alpine climate, no advanced cases being admitted. The contra-indications are much the same as at Arosa (which see). Davos, being 300 metres lower than Arosa, has a correspondingly less stimulating climate; the summer is finer, and is better suited to patients with chronic anæmia. The neighbourhood has been carefully mapped out in quarter-hour walking distances, also showing the steepness of the paths.

Treatment is mainly on the lines indicated by Dr. Brehmer. There are fresh-air galleries 80 metres in length; but patients who are not febrile mostly take a good deal of exercise. They spend as a rule from eight to ten hours a day out of doors. Hydrotherapy is applied in the form of dry and wet rubbings, and in about 25 per cent. as a divided douche, applied by the physician himself. Respiratory exercises with Gifford's apparatus are employed where the lung is little affected. Dr. Turban uses tuberculin where patients are willing to have it, and where the case is suitable. The *sputa* are thrown down the water-closets. The spittoons and spitting flasks are disinfected with solution of corrosive sublimate and sodium chloride. Linen is disinfected by boiling; rooms by being washed with soap and water.

The sanatorium belongs to a company, but is under the control of Dr. Turban and his assistant, Dr. Wunsch. The charges for single rooms are from 2 to 6 frs., for double-bedded rooms 5 to 7 frs. Board is 11 frs., including medical attendance, baths, douches, rubbings, attendance, heating and lighting. There is an entrance fee of 20 frs. for all (excepting servants) who stay more than fourteen days. Other extras are beverages and drugs. There is a reduction in the case of medical men and their families, children and servants; and in summer time 1 fr. per day is deducted from the charges for bedrooms.

Davos is reached by rail from Landquart, which is in connection with Basel and other centres.

THE NEW DAVOS SANATORIUM,

which was recently opened under Dr. Dannegger, is now under the care of Dr. Philippi, formerly at Dr. Turban's sanatorium. It is situated in the English quarter, on the boundary between Davos Platz and Dorf, on the highest part of the road, where it is surrounded by open meadows free from other houses.

From its position it receives a large proportion of sunshine. It has a southerly aspect, and consists of a stone building erected in 1896-7, and a chàlet in Swiss style, both connected by a covered walk with the fresh-air gallery, which is over 60 feet long. The main block, which has a basement ground floor, and four upper floors, on an approximately square base, contains in the basement the kitchen department and heating apparatus. In the ground floor the eastern side is occupied by the dining saloon, which is 12·80 × 4·25 metres (42 ft. × 13 ft. 9 in.). To the south are the drawing room and reading room ; the entrance is on the north side, as well as the water-closets and office and porter's room ; the staircase at the western end of the central corridor. The consulting and waiting rooms are on the first floor over the office. The patients' bedrooms are placed on all sides of the house, but mostly to the south and east. Many have balconies. They all have wood-panelled walls and linoleum-covered floors ; electric lighting, and heating by low-pressure steam pipes. The doors and windows are double, the latter having an upper hinged pane. The water-closets are of a modern pattern, with abundant water flush. The water supply is of excellent quality. In addition to bath rooms in both buildings, a douche room has been fitted up.

The chàlet, which has two upper floors, contains twelve rooms, ten of which are patients' bedrooms. These have parquet floors, but are in other respects like those in the main building. Dr. Philippi sleeps in this part, and has arranged a small laboratory above his room.

The patients admitted, who number forty, or with two in some of the rooms, fifty-five, are mainly consumptives in the more hopeful stages, although others who suffer from anæmia, scrofula, pleurisy, asthma, or bronchial catarrh are not refused. The treatment is after Brehmer's and Dettweiler's methods. Hydrotherapy consists of rubbing, packs and douches, applied before the first breakfast. The douches, which are only applied in suitable cases, last but a few seconds, while the water is cooled by a special mixing apparatus from 68° F. to 47° or 50°. The *meals* consist of a first breakfast at 7·30 to 9 (half an hour earlier in summer), second breakfast at 10·30, dinner of five courses at 1, a light meal at 4, supper of four courses at 7, and extra milk at 9. The milk comes from cows under veterinary supervision. Patients rest from 12 till 1, from 2 till 4, and from 5·30 till 7, as well as after supper. During the afternoon strict silence is observed, to allow those who need it to sleep. The couches have removable horsehair cushions. The patients who are able go for walks before the second breakfast in the morning, and before supper. Creosote preparations and tuberculin T.R. are administered in certain cases. Respiratory gymnastics are only employed in pleuritic adhesions and the like, where no crepitations are audible. In suitable cases patients go up the mountain side through the woods belonging to the Kurverein, about five minutes' walk from the establishment. The place is very thoroughly ventilated day and night. Rooms are rubbed down daily with a damp cloth, disinfected on departure with Schering's apparatus (formalin).

The charges are 10 francs per day for inclusive board and treatment, with 1 to 6 francs extra for the room. Where two sleep together, 4 to 7 francs are charged for the room. Patients buy their own woollen bedcover at cost price, and bring their own rugs, furs, and a new pouch for their napkin ; a footbag reaching up to the chest should be bought in Davos.

The sanatorium is on the telephone, and close to post

and telegraph office. It is ten minutes from the station at Davos Dorf, twenty minutes from that at Davos Platz.

THE MAISON DE DIACONESSES,

founded in 1882, receives advanced cases which are not suitable for treatment in hotels and not admissible to sanatoria. Their charges are a little below those of the hotels.

THE VILLA PRAVIGNAN

is a similar institution exclusively for Roman Catholics, and under the charge of a sisterhood.

THE LEYSIN SANATORIUM,

Leysin, which has the highest parsonage in the Canton de Vaud, is situated above the lower part of the Vallée des Ormonts, on a wide plateau 1264 metres (4150 ft.) above the sea-level, at the foot of the Tours d'Ai. The valley, which has long had a reputation for the cure of scrofulous children, runs from north-east to south-west, the sanatorium being on its northern side, at an altitude of 1450 metres (4760 ft.), which is above the usual fog and cloud level for this part of the Alps. The soil is dry and calcareous, the climate resembling that of other Alpine resorts, with rather more sun than Davos.

The establishment is protected to the north and west by pine woods and mountains, and by high peaks four or five miles farther east. Its aspect is south-south-easterly. It has a fine view across the valley to the snow mountains, of which the most conspicuous is the Dent du Midi, with Mont Blanc in the extreme distance, while Aigle and St. Moritz are visible in the Rhone valley. The forest behind, which extends 300 metres above the sanatorium, has numerous paths, some steep, some nearly level, with sheltered kiosks and seats in abundance.

The sanatorium (fig. 35), which is of stone, consists of five storeys and a basement served by a lift, and contains 110 rooms for patients, of which eighty are on the southern side.

Each of these is at least three metres high, and is provided
with a chimney for ventilation, and with double windows,
the tops of which can be separately opened. There are
roomy balconies in front of most of these. The kitchens,
store rooms, servants' apartments, and staircases are on
the north side. The dining saloon, drawing room, winter
garden, and glazed galleries for exercise indoors, are on the
ground floor and basement. There is also a large *liegehalle*, thirty metres long, on the eastern side of the building, and another is being made farther off. The second
floor communicates with the forest by means of a foot-
bridge. The building is heated by low-pressure steam, a
few rooms also having independent fireplaces. Lighting is
by electricity. Furniture is simple and readily cleansable.
There is an abundant water supply from the Tours d'Ai, by
an aqueduct from a height of 125 metres (410 ft.) above the
establishment. There is also a large *dépendance* (fig. 36).

Patients of all stages are admitted, provided there is any
probability of their benefiting by the climate. Treatment
is on the usual lines, but patients take much rest in the
open air and only a moderate amount of exercise, as at
Falkenstein. They spend on an average ten hours out of
doors. Respiratory gymnastics are not employed. Three
meals a day are provided. Not much alcohol is given, and
this mainly in the form of wine. *Sputa* are received
exclusively into Dettweiler's flasks and spittoons. The
former are disinfected with lysol, the latter with corrosive
sublimate solution, a special male nurse attending to this
every day. The skin is stimulated daily by dry frictions,
occasionally by rubbing with cold, fresh, or salt water. The
douche is not employed. As regards amusements, skating
is recommended for those who are fit for it. There are
two skating rinks. Professional musicians and conjurers
are not admitted; but a café casino has been built at a
short distance, which is of very doubtful advantage.
Smoking is not encouraged, but is allowed in the billiard
room.

FIG. 35.—THE LEYSIN SANATORIUM.

[*Page* 293.

FIG. 36.—THE HÔTEL DU MONT BLANC, LEYSIN.

[*Page* 294.

The building is connected with the telephone and telegraph systems. Two rooms are used for regular religious services, which can also be found in the village for both Protestants and Catholics. Leysin is connected with Aigle by a good road. Diligences run at least twice a day each way and take about three hours. A railway line is also to be made.

The charges are 9 to 15 francs per diem, including service, light and warming. Extras are medical fees, massage, baths and douches, private sitting rooms, meals in private rooms, open fires in the patients' rooms, and a small charge for the use of the open-air gallery. Special diet is charged à la carte. Reductions are made for families, children, and servants.

The sanatorium was built in August, 1892, by a company, and is managed by Dr. Exchaquet, with the help of a managing director.

THE MONTANA SANATORIUM.

Montana is a small village in the Swiss Valais about three hours by road from Sierre, which is itself half an hour by rail from Sion. Some 300 metres above the village of Montana, and 1515 metres (4970 ft.) above the sea-level, is the Hotel du Parc, which until 1st October, 1897, was visited mainly by tourists, but was then reorganised by Dr. Stephani to suit the needs of consumptive patients. The present proprietors are M. Louis Antille and M. Michel Zufferey, the latter of whom also owns a flourishing hotel at Sierre; but Dr. Stephani is endeavouring to form a company to buy the hotel, in which case it would be used exclusively as a sanatorium for consumptives, instead of (as at present) admitting tourists as well as invalids.

The hotel stands on a plateau of over 4 kilometres square, consisting mainly of meadows, the rest being pine woods together with five small lakes. The surrounding country is very mountainous, and mostly covered with pine woods. The Wildstrubel mountain protects the sanatorium to the north, the pine woods belonging to the hotel to the east. The most frequent winds, which are seldom strong, are from the east; those from the west, on which side there is less protection, being rare. There are 4 hectares (9¾ acres) of more or less level wooded land to the north-east of the sanatorium. The soil is schistose, covered with sand.

The climate is very dry, especially in winter, when the hygrometer often records 10 per cent. Fogs are very rare. The winter minimum in 1897-8 was − 14° C. at night, the maximum shade temperature in January, 1898, being + 14·5° C., that in the sun + 47° C.

The sanatorium (fig. 37) faces south on a southerly slope, with an extensive view over the Simplon, Mont Cervin, Mont Blanc, etc., and receives eight hours' sunshine even on the shortest days of the year. The establishment consists of a single building in Swiss style, with high pitched over-hanging roofs, balconies and verandahs; the basement and ground floor and walls of corridors and staircases are built of stone, the rest being of wood. The basement contains the kitchen at the north-east angle, the store rooms, linen room, billiard room, etc. On the ground floor are a large central hall, two drawing rooms, dining saloon, office and two bedrooms for patients. Above this are four floors of bedrooms, the three lower containing each six rooms to the north with areas of 13 to 17 square metres (140 to 188 sq. ft.), and nine rooms to the south with areas of 16½ to 25 square metres (178 to 269 sq. ft.), a corridor running between the two sets of rooms. The fourth floor has one room to the south for visitors, the rest being for the staff. The principal staircase, which is at the back, has leading off it separate water-closets for ladies and gentlemen with water flush on every floor, and on the ground floor and first floor baths and douche rooms. There is a separate staircase for servants at the north-east corner of the building, one room on each floor being for servants. There is no lift. In front of the house is a large glazed verandah, and above it large

Fig. 37.—Montana Sanatorium, Rhone, Switzerland.   [*Face page* 296.

balconies for the first-floor bedrooms. Most of the other
rooms on the south side also have balconies. There are
altogether four galleries on the south side for rest in the
fresh air. There is electric light throughout, and central
low-pressure steam heating. There are very few chimneys,
and ordinary windows, without any special ventilators.
Every window has a wind screen or sun blind. The walls
are papered on the ground floor, elsewhere of pine wood;
those of corridors and staircases being lime-whited. The
floors are of pinewood on most floors, of hardwood on the
ground floor, of red bricks in the corridors. The corners
are not rounded. There are carpets and curtains. The
sewers end on a special meadow 300 metres below the
hotel. The drinking water is from springs 1 kilometre
($\frac{3}{5}$ mile) from the establishment, and conducted to it by a
subterranean conduit.

There is room in the building for fifty patients, who are
received all the year round. As many as eighty tourists
have in former times been lodged in the place. Patients
are received at every stage of illness, although as far as
possible only curable cases are accepted. There is a trained
nurse from 1st October to 1st May.

The treatment is mainly hygienic, by good food and a
regulated life in the fresh air. Three *meals* a day are provided, with plenty of good milk. Where purely hygienic
means do not appear to suffice, Dr. Stephani prescribes cod-
liver oil, phosphates, creosote, inhalations of eucalyptol or
menthol, or any other indicated remedy. Febrile patients
are not allowed to indulge in tobogganing, skating, rowing,
walking, riding, or billiards; but these and other amusements, such as croquet and lawn tennis, are open to such as
are fit for them.

Sputa are exclusively received into spittoons in the rooms,
or into Dettweiler's flasks, which the patients buy, both
being daily disinfected together with their contents by boiling. Linen is boiled in the establishment; bedding disinfected by a Zurich steam disinfector. Rooms are disinfected

by formo-chloral under pressure (Trillat's apparatus) ; the corridors are lime-washed.

The charges for rooms, board and attendance vary from 8 to 13½ frs. per diem. Medical fees are from 30 to 50 frs. per month ; the use of the fresh-air gallery, etc., 3 frs. per week ; in winter 1 fr. extra is charged for heating. Other extras are baths, douches, and meals taken in the bedroom.

The hotel is connected by telephone with the telegraph and telephone office at Sierre. Dr. Stephani, who was formerly at Leysin, has no medical assistant.

THE WEISSENBURG CURHAUS,

in the Simmenthal, is also frequented by consumptive patients amongst others ; but is scarcely to be regarded as a sanatorium for such patients. It stands 890 metres above the sea-level, and has a mild mountain climate with great relative humidity. It is mainly frequented for its thermal, sulphated and saline waters.

FIG. 37*.—THE BASEL SANATORIUM, DAVOS.

[Face page 289.

CHAPTER XLIV.

THERE are several sanatoria for consumptives of small means in Switzerland. The earliest of these was the one at Heiligenschwendi near Thun ; there are also sanatoria at Davos for the inhabitants of Basel, and for those of Dutch nationality; an English invalids' home, and two institutions (Diakonissen-Haus and Villa Pravignan) where seriously affected patients are nursed at a lower charge than in an hotel or ordinary sanatorium. At St. Moritz there is also an English home. Sanatoria are being built at Stachelberg for Glarus, and on the Faltigberg for Zurich : and others are projected at Leysin, Montana, Basel town, and other places. It is expected that in a year or two there will be eight or ten popular sanatoria for consumptives on Swiss soil. There is also an institution for scrofulous children at Aegeri near Zurich. For table of altitudes see p. 283.

THE BASEL SANATORIUM "IN DER STILLE,"

at Davos Dorf, was founded by the *Gemeinnützige Gesell-schaft* of Basel, on the initiative of the medical profession, and with the help of public subscriptions. This society in 1893 formed a committee to consider the advisability of such an institution for the working and the poorer middle classes, and to decide upon a suitable site. After searching the Jura, they came to the conclusion that the Engadine was preferable, and a site was obtained at the foot of the Seehorn, near the entrance to the Fluelathal, 1600 metres (5250 ft.) above the sea-level. The sanatorium is in a very sheltered situation, with a south-westerly aspect, being protected to the east and west by pine woods. The ground

(299)

amounts to 2 hectares (4¾ acres), half of which was given by the Huntsmen's Society. The building was begun in June, 1895, but, work being interrupted by the winter snow, was not completed until December, 1896.

In order to expose as little surface as possible to the cold air, the sanatorium was built on a concentrated plan, instead of the extended one-sided arrangement common in the lowlands. It was also placed as far as possible to the north-west of the ground, to allow of future enlargement, and the administrative portion prepared for 100 patients, whereas the bedrooms would only at first accommodate seventy. It consists of an oblong block 48·2 metres long, 20 deep, and 25 high, with a projecting wing behind the south-east end, to accommodate kitchen and dining room.

The main block consists of a basement, ground floor, and four upper floors, including the attic storey; the kitchen pavilion only goes as high as the second floor, and has a flat roof, which can be reached from the third-floor corridor, whereas the main block has a rapidly sloping roof to shed the snow, with lozenge-shaped metal tiles, gutters which do not project, and chimneys which come up to the roof-ridge. The basement (fig. 38) contains in the centre the heating apparatus; next this the wine cellar, laundry, disinfecting oven, workroom, and store rooms; and in the wing, the kitchen department and servants' hall. On the ground floor are the office and manager's rooms, porter's room, waiting and consulting rooms, laboratory, bath and douche rooms, a large vestibule on the north side, and two large common rooms on the south side, for men and women respectively. At the junction with the wing is another large common room, and in the wing itself the large dining saloon over the kitchen. Along the southern side extends the open verandah with room for fifty couches. The floor of this verandah is sunk a few steps; it is of wood over cement, and rests on a projecting part of the basement. The first floor (fig. 39) contains ten patients' bedrooms on the south side and the chief medical officer's rooms at the

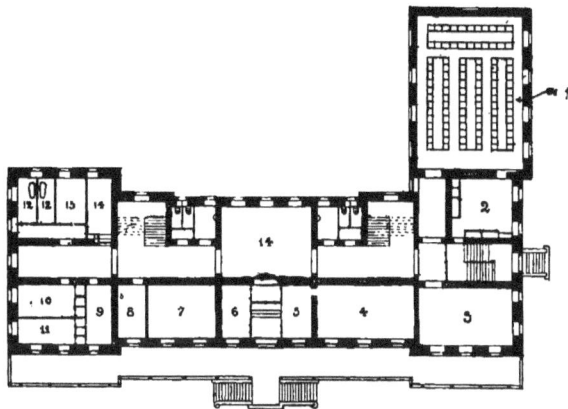

GROUND FLOOR

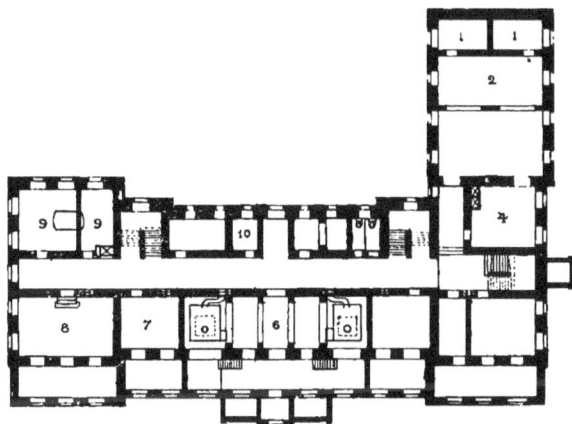

BASEMENT

Fig. 38.—The Basel Sanatorium at Davos.—Ground Floor and Basement.

Ground Floor :—

1. Dining Saloon.	6. Porter's Room.	11. Laboratory.
2. Serving Room.	7. Women's Sitting Room.	12. Bathrooms.
3. Common Sitting Room.	8. Nurses' Room.	13. Dressing Room.
4. Men's Sitting Room.	9. Waiting Room.	14. Douche Room.
5. Office.	10. Consulting Room.	14 (centre). Vestibule.

Basement :—

1. Larders.	4. Servants' Hall.	7. Wine Cellar.
2. Scullery.	S.E. corner, behind Balcony, Workroom.	8. Laundry.
3. Kitchen.	6. Heating Apparatus.	9. Disinfector.
	10. Mortuary.	

[*Face page* 300.

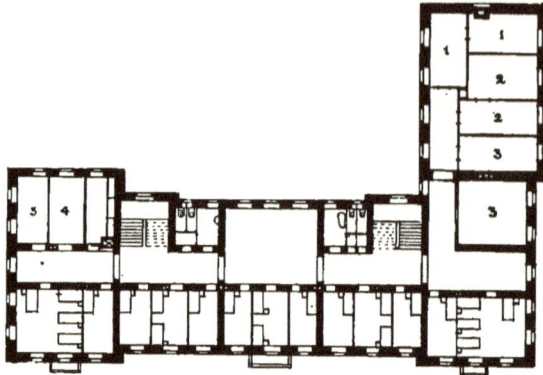

SECOND FLOOR

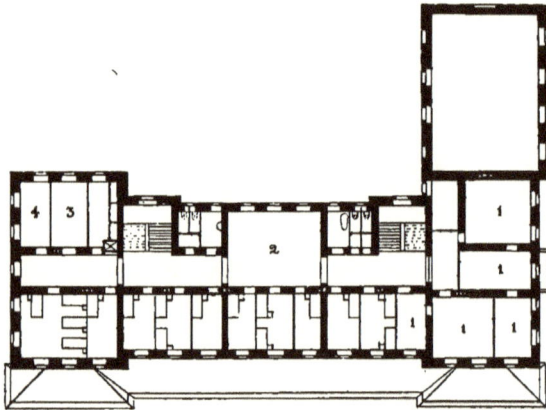

FIRST FLOOR

FIG. 39.—THE BASEL SANATORIUM AT DAVOS.—UPPER FLOORS.

First Floor :—

1. Medical Officer's Quarters.
2. Vestibule.
3. Nurses' Room.
4. Cloak Room.

Second Floor :—

1. Manager's Quarters.
2. Assistant Medical Officer's Quarters.
3. Reserve Rooms.
4. Nurses' Room.
5. Cloak Room.
Centre of North side, Vestibule.

[*Face page* 301.

junction with the wing. In the latter there is no accommodation on this floor, owing to the greater height of the dining saloon.

In the second floor are thirteen patients' rooms on the south side, the assistant medical officer and manager being accommodated over the dining saloon. The next floor is similar ; in the attic floor is a large drying room, an ironing room, and a clothes lift to the laundry in the basement ; also a workshop for the use of the patients at times. On every lower floor there is a large vestibule on the north side for the use of the patients, and a bath room for such as cannot safely use the ones in the ground floor ; the closets, lavatories, attendants' and nurses' rooms and main staircases being also at the back.

There are altogether seventy beds, seven rooms containing four each, ten rooms two each, and twenty-two being single-bedded. The cubic space is 40 cubic metres (1413 cub. ft.) in the bedrooms, and at least 28 cubic metres elsewhere. The height of the ground floor is 3·25 metres (10 ft. 8 in.) ; of the first, 2·90 metres (9 ft. 6 in.) ; second, 2·85 metres (9 ft. 4 in.) ; third, 2·80 metres (9 ft. 2 in.). The dining room is 5 metres (16 ft. 5 in.) high. It should be remembered that in an Alpine climate ventilation is much more rapid than elsewhere.

The sanatorium is built of stone, with an inner lining of hollow tiles ; stairs and foundations of cement ; main staircases of oak: floors in the patients' quarters of oak or boxwood, or else of pine covered with linoleum ; in the kitchen, "terrazzo" ; elsewhere in the basement of cement. The window frames are of unpainted red pinewood ; the windows are double, with upper panes which open inwards together. The inner walls are smooth and free from mouldings, with rounded corners and angles, and covered with oil-coated and washable linen paper. In the laboratory, douche rooms and bath rooms there are enamel-painted walls ; in the common rooms and dining saloon breast-high wooden panelling.

The furniture is of plain lacquered pinewood. The sanatorium is lighted by electricity, and heated by low-pressure steam. The heaters are placed near the windows and outer walls, and have no dust-catching covers. The ventilation is by open windows alone, excepting in the kitchen, dining room, and laundry, which have additional contrivances. The water-closets pass their contents into "Fosses Mouras" on their way into the Fluela stream. These fosses are iron or cement vessels filled with water and containing little or no air. The entering drain is in the upper third; the exit about the middle of the side. Under these circumstances the contents are disintegrated by fermentation, and the effluent is said to be nearly colourless and inodorous. The water supply is that of Davos. A cistern, 19 metres (62 ft.) above the roof-level and containing 25 cubic metres (88 cub. ft.), supplies the whole establishment.

The building has cost nearly £14,000 up to the present, not including the cost of the site. The third floor is also not yet completely furnished. Only forty-five beds were at first opened; there are now sixty-one available out of seventy.

Only phthisical patients, and those threatened with consumption, are admitted; and no hopeless cases are admissible. The usual payment is 5 frs. per diem; but those who on investigation prove to be unable to pay as much are admitted for 3 frs. per diem. Those who pay the full charge have first choice of single rooms. Sick societies and hospitals that subscribe towards the funds of the sanatorium can send patients at the rate of 2 frs. per diem. Washing is charged extra, excepting to the occupants of the four-bedded rooms. There are four examining physicians in Basel, who receive applications signed by the family doctor and examine the patients to determine their fitness for admission. The *Basler Hilfsverein für Brustkranke*, founded in autumn, 1896, receives subscriptions to help suitable poor patients and their families; in some cases paying railway expenses, in others part of the cost of treatment. They also

endeavour to educate the public in the prevention of tuberculosis, and to find suitable employment for convalescents. A
certain number of beds are reserved for the inhabitants of
Basel town and Basel country respectively, others being
admitted irrespective of habitat or nationality if there
should be room.

Patients undertake to stay at least thirteen weeks if
thought advisable by the medical officer. During the first
seventeen months (" first year ") 185 patients with 17,553
days of treatment were admitted, making an average stay
of 104 days. Some only stayed three or four weeks, some
over eight months. Both sexes are admitted, the men
being in the majority. They take their meals in common;
but are separately treated and housed in every other
respect. During the first (nominal) year the cost per head
was 4·09 frs. per diem ; or, without the expenses of
administration in Basel town, about 2·50 frs. The average
receipts were 2·80 frs. The establishment is free from debt,
and has acquired an adjoining wood at a low price.

The treatment resembles that in other Alpine sanatoria.
Patients stay in the open air from 7 A.M. till 9·30 P.M., excepting meal times. Graduated exercise or rest is observed
according to the doctor's prescriptions. Five *meals* are provided, with milk at bedtime. Wine is only given on special
occasions or upon the doctor's order, the money being more
usefully devoted to improving the food.

Sputa are received exclusively into Dettweiler's flasks,
ordinary spittoons not being used. Linen is put by the
patients into a bag, which is emptied into a metal chest and
washed every fourteen days. Bed linen is changed every
fortnight ; towels every week ; blankets, etc., at longer
intervals. If patients die or are bedridden before leaving
the sanatorium the bed and bedding are disinfected. The
disinfection is by steam at 105° to 108°C. Rooms and
corridors are daily cleansed with a damp cloth, and at
longer intervals with soft soap and water.

Out of those treated during the first year, 20·54 per cent.

were severely affected, 32·98 per cent. moderately affected,
27·57 per cent. were in an early stage,18·91 per cent. in a later
stage but only slightly affected. Of all these collectively
25·38 per cent. left apparently cured, 42·31 per cent.
decidedly better, 23·10 per cent. somewhat better, 3·84 per
cent. stationary, 3·84 per cent. worse, 1·53 per cent. died.
This gives a total improvement of nearly 91 per cent.

The sanatorium is a quarter of an hour from the railway station of Davos Dorf. Dr. Kündig is chief medical
officer ; he is aided by Dr. Buser.

It is intended to establish another sanatorium of
sixty beds in the neighbourhood of Basel town, in the
Budenholz, for such as cannot be usefully sent to Davos.
Cliniques are also to be established in connection with
the City Hospital.

For climate, etc., see p. 287.

THE DUTCH SANATORIUM AT DAVOS

for poor people was founded in 1897 by a private society,
and is maintained chiefly by voluntary subscriptions from
residents in Holland, including the Queen, and is intended
exclusively for consumptives of both sexes of Dutch nationality. It is situated 1560 metres (5120 ft.) above the sea-
level, in a private house which has been recently leased for
the purpose. It has no private garden, but its patients
have the right to use the Curgarten and parks and woods
of the place.

The building is on the hillside, sheltered from the north
and west by mountains, and receives on an average seven to
eight hours' sunshine in the winter. It was built about
ten years ago in the usual Swiss style, and consists of a
basement, three upper floors and attics. The bedrooms are
200 to 250 square metres in area, with one to three beds
each ; those with three having a cubic capacity of 400 to
500 cubic metres. The walls are of varnished wood, or
covered with washable paper. Each has a closed stove,
and wooden furniture, no stuffed furniture being used and

but few curtains. The floors are everywhere covered with linoleum. The lighting is by electricity. The windows are "French windows," with an extra hinged pane above, which opens inwards. There is also a dining room, drawing room with piano, and a suitable bath room. Most of the rooms on the south side have balconies. In addition to this there are three large verandahs at the western side of the house, of which the lower is glazed, the upper ones being provided with solid roofs and thick curtains to draw down on the windy side. The water-closets have a good flush. The sewage is carried into the Landwasser through the town sewers.

There are thirty beds. Patients pay 4 francs per diem, including board, lodging and medical attendance.

They spend, as a rule, from 9 A.M. till 9·30 P.M. in the open air, and an hour or more in walking abroad, morning and afternoon. The cold wet rub is frequently employed, the cold douche but seldom. Cod-liver oil is exceptionally given, but no "specifics," and no stimulants. Random spitting is strictly forbidden, as also the use of handkerchiefs for the *sputa*, which are received into the spitting-flasks and spittoons, and emptied into the water-closet. Linen is disinfected in the public steam disinfector. Rooms are washed with soda solution, and in the case of death also disinfected with formalin vapour.

There is a trained nurse in the house. Dr. A. Schnöller is the resident medical officer. The sanatorium is to be open all the year round. There is some talk of affiliating the institution to the recently formed public society in Holland.

During the first four months (1st Sept., 1897, till 1st Jan., 1898) thirty-five patients were received, and a profit of about 550 francs was made, which was used in buying more furniture. The sanatorium cost 12,500 francs for furniture and installation. When sufficient funds are forthcoming, the society intend to enlarge the institution to fifty beds, and eventually to build a sanatorium of their own.

20

THE DAVOS INVALIDS' HOME

is intended for ladies and gentlemen of limited means who are in need of treatment in an Alpine climate. It was founded in 1884 by three English ladies (Mrs. Lord and the Misses Crothers), and carried on by them for eleven years, until Mrs. Lord's ill-health compelled them to relinquish the work. It has since been carried on by a committee with the aid of public subscriptions. The institution is managed by an English lady, and without any denominational restrictions. The charges are 4 to 4·50 frs. per diem, including board and residence, medical attendance and nursing, the only extras being medicine and personal washing. The hon. secretaries are Mr. Arthur Herbert, 6 Finch Lane, London, E.C., and Dr. Wm. Ewart.

A *benevolent society* exists at Davos to secretly assist those whose funds are exhausted before the end of their course of treatment in one of the paying sanatoria.

THE ST. MORITZ AID FUND

is also intended to provide those of limited means with treatment at an Alpine health resort. The president is Princess Christian. Applications may be made in the first instance to Lady Jeune, 79 Harley Street, London, W., or to Lady Bancroft, 18 Berkeley Square, W.

THE HEILIGENSCHWENDI SANATORIUM

was opened in August, 1895, to commemorate the 600th anniversary of the Swiss Confederacy and the 700th of the City of Bern, mainly on the initiative of various members of the medical profession. It stands, about 1½ hours' walk from Thun, on an elevation overlooking the Lake of Thun, 1160 metres (3800 ft.) above the sea-level, with a lovely view across the lake to the Niesen, Jungfrau and the Alps. The situation is sheltered, and not subject to the *Föhn*. To the north is the Blume, to the west are wooded hills. There is a fairly large garden in front, and although there is very

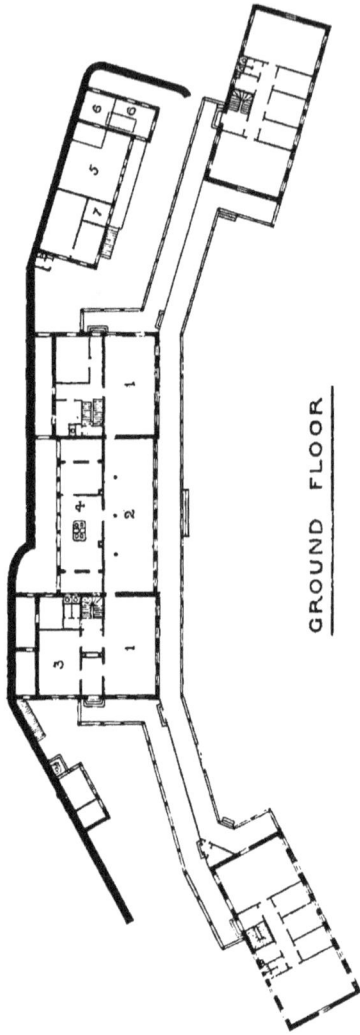

GROUND FLOOR

Fig. 40.—The Heiligenschwendi Sanatorium.

1. Sitting Rooms.   2. Dining Saloons.   3. Waiting and Visitors' Room.   4. Kitchen.   5. Laundry.
6. Disinfector.   7. Stable.

[Face page 307.

little ground behind the institution, patients are able to use the public woods close by.

The sanatorium, which faces south-south-west, consists of a central building (fig. 40) united by covered corridors with two slightly projecting lateral pavilions. Behind the eastern pavilion is a separate *dépendance*. In front of the central block and the connecting corridors are the fresh-air galleries, which were originally open, but have since been glazed. In the central block is the dining saloon, and on either side of it a common room. The former is for both sexes, but in other respects they are kept apart. The northern part of the central block contains the kitchen, etc. On the first floor are a few single-bedded rooms to the south, and over them a number of attic rooms. Both lateral pavilions have on the ground floor two large rooms with eight beds each and windows on three sides, and a few smaller rooms for one or two beds apiece, and the same on the first floor. The *dépendance* contains the manager's and doctor's quarters. There is also an outbuilding behind the western corridor with laundry, disinfector, stables, etc. The inner walls are painted with glossy oil paint; the floors are of simple washable parquet. The building is heated by low-pressure hot water pipes, and lighted by petroleum, although electrical lighting is to be introduced in future. There are no special ventilating contrivances; but the windows are very large, reaching down to the ground, the lower part being guarded with a grating. There are good baths, a good water supply, and "Unitas" water-closets, the house drain leading to a cesspool with an overflow. The furniture is simple. The lounge chairs are in two pieces and can be used as armchair or couch.

There are at present fifty-two beds, including four for children; but there will eventually be 100 or 120 in four separate blocks. None but consumptive patients are admitted, and no serious cases are supposed to be admitted; but a good many such have been sent in notwithstanding, so that on an average one-third have been in bed, and nineteen died out of 186 in 1895-6. Patients on arrival have a bath, during which their clothes are disinfected. There is a fine of 1 fr. for spitting on the ground. Patients buy their own Dettweiler's flask, which is disinfected with carbolic solution. In the bedrooms they may use *spitcups* of glass, which are emptied into the water-closet. Special handkerchiefs are provided by the institution for mouth-wiping in bed. These are collected twice a week and disinfected, with the other linen, by boiling.

Patients send a *menu* daily to the manager, who provides accordingly if it seems advisable. They have about two litres of milk per day. They take walks according to the advice of the medical officer, Dr. Glaser of Steffisberg, who comes over twice or three times a week, and is in communication with the building by telephone. In winter some are allowed to sledge on the mountain side. Patients do a little work for the establishment, both out of doors and in the kitchen and elsewhere. So far there has been no hæmoptysis in consequence of this regulation. Some difficulty has been experienced in entertaining the men, games and reading being insufficient, so that the erection of a small workshop is contemplated.

The staff consists of a manager, a matron, two nurses, two maids, a helping maid, and a man to attend to the disinfection and heating. Thanks to good means of communication, this staff has been found quite sufficient. There is no resident medical officer.

Patients stay at least two months. Up to 1st July, 1896, there were 186 patients, ninety-eight being men. Their days of treatment amounted to 12,843, or an average of forty days; but most of the patients stay three months. The average daily number was forty; but the building is now nearly always full. The daily cost amounts to 1·90 frs. Patients pay from 1½ to 4 frs., according to their means. Those who come from subscribers, parishes, hospitals or societies in the canton of Bern have the preference in admission. The total cost of the building was under

200,000 frs. without the land, or, with road-making, 280,000 frs. This is about £216 per bed.

THE ZURICH SANATORIUM

is being built near Wald, on the Faltigberg, a little north of the southern end of the lake of Zurich, 900 metres (2950 ft.) above the sea-level. The ground amounts to 30 hectares (74 acres), with woods and meadows. There will be from ninety-two to 100 beds.

The building will have a southerly aspect, and will have an administrative block in the centre, and a three-storey pavilion on each side, for men and women respectively, with open-air corridors between. It will be opened on 1st October, 1898.[1]

THE SANATORIUM OF BRAUNWALD,

for the canton of Glarus and other neighbouring cantons, has just been opened on an elevated terrace near the baths of Stachelberg, in the valley of the Linth, 600 metres above the stream and 1160 metres (3805 ft.) above the sea-level. Although intended for the poor, it will also receive paying patients.

Surrounded by woods and meadows, it has a south-south-westerly aspect, and is well sheltered by mountains from north-east winds. It has a dining room, work room, fresh-air gallery, and rooms for gymnastics and games in bad weather, as well as beds for thirty patients, from one to three in each bedroom. There are two windows to each bedroom, and good ventilation by means of hinged upper window panes, about 30 cubic metres (1060 cub. ft.) air space being allowed per head. The walls are wainscoted to a man's height, and all corners rounded. The heating is by hot water pipes, the lighting by electricity. Furniture is simple, of varnished wood. A trained nurse is in charge of the establishment, which

---

[1] Liebe, *loc. cit.*; *Correspondenz Blatt für Schweizer Aerzte*, 1st July, 1898.

has no resident medical officer, but is visited two or three times a week by a doctor from Glarus. There is a good water supply. A plentiful but simple dietary is provided. The douche is used for convalescents, but no cod-liver oil, "specifics," or counter-irritants. *Sputa* are disinfected by lysol, linen by boiling with soda ley, rooms with corrosive sublimate.

Patients from the canton of Glarus pay 2½ frs. per day, or 4 frs. in a single-bedded room. The sanatorium is 1¼ hours by road from Linthal and Rüti; Linthal is 3½ hours by rail from Glarus. According to Kuthy, another sanatorium will later on be built on the same site by a neighbouring canton, when a resident medical officer will be appointed.

THE AEGERI SANATORIUM.

A sanatorium for scrofulous children has existed since 1885 at Aegeri near Zurich, at an altitude of 820 metres (2690 ft.) above the sea-level.[1]

---

[1] Liebe, *Hyg. Rundschau*, 1896, No. 14.

CHAPTER XLV.

SOME other countries, both in Europe and elsewhere, are beginning to bestir themselves in the campaign against tuberculosis, although up to the present they have done but little in establishing sanatoria.

BELGIUM.

At the Congress of Hygiene and Medical Climatology, held in Brussels in August, 1897, a resolution was unanimously passed that a league against tuberculosis should be formed on the model of the French *Ligue contre la Tuberculose*. Another resolution was passed urging the foundation in Belgium of sanatoria for consumptives of the poorer classes. Several seaside convalescent homes for children already exist—such as the one at Middelkerke with 300 beds, that at Venduyne with 200, and that at Ostende. Another convalescent home for children exists at Esneux, about half an hour's rail from Liège, and at Bockryck-Genck-lez-Hasselt and Bonsecours for adults, while a third is to be built at Crainhem-Tervueren. All three are inland at a low altitude. A hydropathic establishment like the French one at Divonne is to be erected at Venaimont in the Belgian Ardennes.[1] Some of these might possibly be utilised for the open-air treatment.

BULGARIA

has also a project for the erection of separate sanatoria for consumptives in each district. A medical man has

[1] *Brit. Med. Journ.*, 9th April, 1898.

been sent by the Government on a journey of investigation.[1]

DENMARK.

For some time past consumptives have been boarded out in various farms in the country with satisfactory results. Sanatoria exist at Refsnaes with 130 beds for scrofulous children, and at Heidschloss, near Ploen, in Holstein. Dr. Schepelern is in charge of the former. More recently a society has been formed for the erection of sanatoria for paying consumptive patients, the subscribers binding themselves to receive no more than 4 per cent. interest, all further profit to go towards the maintenance of free beds (Liebe, *Hyg. Rundschau*, November, 1897). Over £10,000 have been subscribed, and it is stated that the first sanatorium will be opened this year.[2]

EGYPT.

Several hotels exist in various parts of Egypt with resident medical officers, and almost entitled to be called sanatoria. They are, however, by no means exclusively devoted to consumptives, and the climate does not permit of an open-air life throughout the year. Among the best known hotels which might safely be visited are two at Helouan, where Dr. Page May lives; Mena House, near the Pyramids (Dr. Bentley); three at Luxor; and one at Assuan.

HOLLAND.

A society for the erection of sanatoria for consumptives of Dutch nationality was formed in 1896 by a number of medical men, with Prof. Reisz, Rector of the University of Utrecht, at their head, and Dr. Saugman as secretary. They succeeded in raising over 300,000 frs. in shares and donations, and the State voted an additional 138,000 frs. (*Rev. de la Tuberculose*, December, 1896). In May, 1897, it was decided to form a number of local sub-committees to

---

[1] *Das Rothe Kreuz*, 1897, No. 16.
[2] Léon Petit, Tuberculosis Congress, Paris, 1898.

raise more money, and to erect two sanatoria, of which one was to be on the coast. In September, 1897, a private society leased a house at Davos and converted it into a sanatorium for thirty patients (see p. 304). Two convalescent homes for children also exist on the coast of Holland at Zaandvoort, near the Hague, and at Wyk van See ; but these are not exclusively for tuberculosis (*Heilst. Corresp.*, January and July, 1898).

ITALY

has about twenty seaside hospitals for tuberculous diseases of the bones and joints, but up to the present no sanatoria for consumptives.

JAPAN.

It is stated that a sanatorium for consumptives is to be erected in Japan under the patronage of H.I.M. the Empress.[1]

PORTUGAL.

At a recent Congress of Hygiene resolutions were passed in favour of the establishment of sanatoria for consumptives, and of the adoption of other measures for the prevention of tuberculosis.

ROUMANIA.

The Council of Hygiene of Bucharest recently appointed a committee to prepare a handbook of advice as to the means of preventing phthisis. The committee has recommended, among other things, that persons suffering from the disease be excluded from workshops and placed in sanatoria where they should be treated and maintained at the cost of the municipality. Steps have already been taken to obtain buildings for transformation into a sanatorium. A special medical inspector is appointed to discover cases of tuberculosis in workshops and factories.[2]

[1] *Das Rothe Kreuz*, 1897, No. 17.
[2] *Brit. Med. Journ.*, 2nd July, 1898.

SPAIN.

Beyond the *Istitute Rachitici* for children, near Milan, under Dr. Panzeri, there are no special institutions for the hygienic treatment of tuberculosis in Spain, and up to the present no sanatoria have been erected there.

CHAPTER XLVI.

BRITISH COLONIAL SANATORIA.

NOTWITHSTANDING the great climatic advantages possessed by several of our colonies, but little has been done up to the present in erecting suitable colonial sanatoria.

SANATORIA IN AUSTRALIA AND NEW ZEALAND.

It is doubtful whether any sanatoria in the strict sense of the word are yet in existence in Australasia for the treatment of consumptives according to Brehmer's and Dettweiler's methods, although there has been some talk of establishing such institutions. There is a sanatorium for consumptives at Echuca, Victoria, with fourteen beds; and the Austin Hospital for incurables at Heidelberg. Victoria has twelve beds for advanced and practically incurable consumptives. There are also homes for consumptives at Parramatta and Thirlmere in N.S. Wales, the latter founded and for some years entirely maintained by the generosity of Captain Goodlet, but lately dependent upon public subscriptions and a Government grant. A scheme was proposed in 1897 to commemorate the Queen's Diamond Jubilee by the establishment of a hospital near Sydney for chronic and complicated cases of phthisis, together with sanatoria in suitable parts of the country, and a hospice for the incurable cases. Up to July, 1898, under £14,000 had been raised for this fund, so that it was decided to devote part of the interest from its investment towards the maintenance of the Thirlmere Home during the next two years.[1]

[1] *Lancet*, 7th July, 1894; *British Medical Journal*, 10th September, 1898; private letters from Dr. J. E. Kyngdon.

(315)

In Queensland "arrangements have been made for the establishment of a ward for consumptives at the hospital at Roma, and also at Dalby. According to Dr. Hardie, Dalby, Roma, and Charleville in S.W. Queensland are admirably suited for the treatment of phthisis during the hot months from October to April, as also a high altitude station, Stanthorpe on the Dividing Range, while Mount Tambourine in S.E. Queensland" is the best for early cases during the cooler months from May to September.[1]

CANADA.

It is proposed to erect a sanatorium for consumptives on *Trembling Mountain*, overlooking the village of Ste. Agathe. The Government intend to set aside sufficient Crown land (100,000 acres) to form a natural park, to be called the Trembling Mountain Park; and it is here that the sanatorium will probably be built, at an elevation of 2500 feet above the sea-level. The district resembles that of the Adirondacks (see p. 84), and is situated among the Laurentians, north of Montreal.

The Muskoka Cottage Sanitarium, for incipient phthisis, has been recently founded in a region abounding in pine trees, about 100 miles north of Toronto and 800 feet above the sea-level. There is shelter from mountains and forests to the north and west, whence come the colder winds in winter. The grounds, which embrace seventy-five acres, are on Lake Muskoka. The present establishment, which accommodates forty patients, consists of a large and well-planned main building, surrounded within easy distances by a number of small cottages.[2]

CAPE COLONY.

A sanatorium, founded by the Right Hon. C. J. Rhodes at Kimberley, in Cape Colony, near the Orange Free State, was opened on 1st September, 1897.

---

[1] *Lancet*, 15th May, 1897.
[2] Pres. Address by Dr. Roddick, at the annual meeting of the Brit. Med. Assoc. at Montreal, 1897.

Kimberley is on the high inland plateau of South Africa, 650 miles from Capetown, with which it is connected by rail.

The climate is dry and sunny : the rainfall being 18 in. per annum, the mean annual humidity 55 per cent. of saturation ; the mean summer temperature (October to March) 72°, the mean winter temperature (April to September) 56°.

The sanatorium is about 4100 feet above the sea-level, about half a mile from the town, on the tram line to Beaconsfield. It stands on seven or eight acres of ground on the brow of a hill, overlooking the town and the neighbouring Free State hills, and has well laid-out tennis and croquet courts.

It contains about thirty bedrooms, several of which can be arranged *en suite* ; large dining room, drawing room, ladies' morning room, smoking and billiard rooms, together with the appointments of a first-class hotel, and excellent sanitary arrangements.

It is intended for early and curable cases of consumption. The charges, which are intended to render the institution self-supporting, are from 15s. to £1 per diem according to room. There are special rates for the week or month.

Applications to the secretary, Kimberley sanatorium, with a banker's reference.[1]

---

[1] *Lancet*, 26th March, 1898.

CHAPTER XLVII.

BRITISH INSTITUTIONS FOR CONSUMPTIVES OF THE POORER CLASSES.

IN reviewing the provision for the poorer classes in this country, it is advisable to remember the definition of a sanatorium given in the first chapter of this book, and to place those institutions in a separate class where the treatment is an indoor treatment, or where the situation or the absence of adequate supervision render the open-air methods of little value. Judged by this standard, the British Isles appear to be rich in urban chest hospitals and in convalescent homes, but poor in sanatoria and nursing homes; rich in institutions for the patching of consumptives, poor in those for effectually mending them.

There are in London four chest hospitals with 645 beds; while in the country, including those in provincial towns and cities, there are twelve hospitals (one in process of erection) and five homes or sanatoria (one being built), with a present total of 517 beds. There are also over 150 *convalescent homes*, which do not exclude consumptives in an early stage, and which have collectively between 7000 and 8000 beds; as well as five *homes for advanced consumptives* (one in London), with altogether 108 beds.

LONDON HOSPITALS FOR CONSUMPTIVES.[1]

	Founded.	Beds.
Brompton Hospital for Consumption and Diseases of the Chest	1841	321
City of London Hospital for Diseases of the Chest, Victoria Park	1848	164
Royal Hospital for Diseases of the Chest, City Road	1814	80
North London Hospital for Consumption and Diseases of the Chest	1860	80

[1] Several of these hospitals have been obliged to close some of their beds temporarily for want of funds.

(318)

For the treatment of *out-patients* there is another special institution in London, the Infirmary for Consumption and Diseases of the Chest, which has no beds. During 1897, 3611 in-patients and nearly 40,000 out-patients were treated in the five London institutions. None of these can be regarded as sanatoria; nor can their out-patient departments be regarded as a satisfactory means of treating the disease, so long as the faulty conditions of daily life are ignored. Consumptives also attend as out-patients at the London general hospitals, and are occasionally admitted for hæmoptysis or some other urgent complication. They are, however, not welcomed in these institutions, and are transferred as soon as possible to a special hospital or poor law infirmary. Consumptives in advanced stages are ineligible for admission to the London general hospitals, and two of these refuse to admit them in any stage. The poor law infirmaries of London and the suburbs have collectively over 13,500 beds for the reception of a variety of more or less chronic diseases amongst the poor. Like the London general and chest hospitals, they have large and well-ventilated wards; but few of them are ideally placed for open-air methods of treatment, which, moreover, their medical officers are far too busy to supervise.

Of the provincial hospitals for consumption the only one where systematic open-air treatment is adopted is, I believe, the one at Craigleith in Scotland,[1] although the treatment is carried out in a modified form at Ventnor and elsewhere. In the small sanatoria for the poor at Cromer and Downham the open-air methods are fully carried out, as far as the absence of a resident medical officer and of extensive grounds will permit. The following is a list of special hospitals and sanatoria in the provinces:—

[1] Unless the National Hospital for Consumption for Ireland carries it out.

COUNTRY HOSPITALS AND SANATORIA FOR CONSUMPTIVES OF THE POORER CLASSES.

ENGLISH.		Founded.	Beds.
Ventnor . .	Royal National Hospital for Consumption and Diseases of the Chest .	1868	134
Bournemouth .	National Sanatorium for Consumption and Diseases of the Chest . .	1855	62
St. Leonards .	Eversfield Hospital and Home for Consumption and Diseases of the Throat and Chest	1891	55
St. Leonards .	Winter Home for Consumptive Girls .		12
Torquay . .	Western Hospital for Patients of Consumptive tendency	1850	40
Worthing . .	Richmond Hospital and Home . .	1891	24
Cromer . .	Fletcher Convalescent Home (Experimental Sanatorium) . . .	1895	(30) 6
Clewer, Windsor	St. Andrews' Hospital for Convalescents and Incurables (Special Ward)	1861	(80) ?
Bowdon, Cheshire	Manchester Hospital for Consumption and Diseases of the Chest . .		50
Liverpool . .	Hospital for Consumption and Diseases of the Chest	1864	44
West Kirby, Cheshire	Convalescent Home for the above .		?
Newcastle-on-Tyne	Northern Counties' Hospital for Consumption and Diseases of the Chest	1878	5
Downham, Norfolk	Miss Walker's Cottage Sanatorium .		5 or 6
SCOTCH.			
Craigleith, Edinburgh	Victoria Hospital for Consumption and Diseases of the Chest . . .	1894	15
Bridge of Weir, Renfrewshire	Hospital for Consumption (being erected)		(38)
IRISH.			
Newcastle, Wicklow	National Hospital for Consumption for Ireland	1896	24
Belfast .	Forster Green Hospital for Consumption and Diseases of the Chest .	1887	40

The number of patients annually relieved in these institutions is over 2100. It should be noted that nearly all our special hospitals and homes for consumptives also admit sufferers from other complaints, such as heart disease, aneurism, pleurisy, pneumonia, bronchitis, and asthma, thereby practically diminishing the available accommodation for phthisical patients. A well-known London physician, indeed, asked me one day in a joke whether I expected to find any consumptive in-patients in the London chest hospitals.

The following is a list of nursing homes for advanced consumptives of the poorer classes:—

HOMES FOR ADVANCED CONSUMPTION.

		Founded.	Beds.
London . .	Home for Consumptive Females .	1863	26
Bournemouth .	Firs Home	1868	20
Torquay . .	Mildmay Consumptive Home . .	1886	10
Ventnor . .	St. Catherine's Home for Patients in Advanced Consumption . . .	1879	12
Cheddar,Somerset	St. Michael's and All Angels' Home for Consumptive Men and Women	1878	40

In all but the last a payment is required of 7s. to 10s. 6d. per week. In the special hospitals admission is sometimes free, at others by payments of various sums up to 17s. per week for the poor. Some also admit patients of a better class at £1 1s. or £2 2s. per week. There are, further, a number of cottage hospitals and large general hospitals in various parts of the country where consumptives are admissible as well as other kinds of patients.

A lengthy description of our urban chest hospitals would be foreign to this book, the purpose of which is to describe the institutions for consumptives that employ the modern open-air methods of treatment. When the atmosphere of London has been purified by the enforcement of stringent smoke acts, it may be possible for all of these hospitals to adopt the open-air methods with benefit to their patients ; but under present circumstances there is only one (the North London C. H.) where such a change would be of much use.

THE.VENTNOR CONSUMPTION HOSPITAL,

or, more correctly, the "Royal National Hospital for Consumption and Diseases of the Chest," was founded by the late Dr. Arthur Hill Hassall in 1869 on the undercliff, less than a mile west of Ventnor. Originally formed of a single block, it now consists of ten, together with a handsome chapel, while an eleventh block is rapidly approaching completion (fig. 41).

The site of this hospital is in many respects an admirable one. Protected to the north by a thickly-wooded slope which rises to the greensand cliffs surmounted by the lofty chalk downs, to the east by a rising patch of undercliff in front of Steephill Castle, and to the west by tall trees, it is separated from the sea by about 300 yards of undulating undercliff, part of which, near the hospital, has been converted into garden and shrubbery. The soil is of sand, the elevation inconsiderable, but sufficient to ensure a dry situation. About twenty-two acres of ground belong to the hospital. In the meadow to the east is an open shelter, and others will probably be added in the course of time.

The climate of the undercliff is known for its mildness. The winters are warmer than in most other English health resorts, and the summers slightly cooler than further inland. There is a large proportion of sunny weather, and a comparatively small number of rainy days in the year. The mean humidity, although greater than that of Bournemouth, Brighton, and Weymouth, and of the stations on the Riviera, is less than that of most British health resorts, the annual mean being 81 per cent., and that of the winter months 84·3 per cent.[1] Ventnor is probably less bracing than the Kentish and east coast seaside resorts, the German hill sanatoria, and some of the inland districts of England.

The separate blocks composing the hospital are of brick and stone, with slated roofs, and are only united by an underground passage which runs the whole length of the

[1] *The Climates and Baths of Great Britain.* London, 1895. Vol. i., p. 212.

FIG. 41.—THE ROYAL NATIONAL HOSPITAL FOR CONSUMPTION AND DISEASES OF THE CHEST, VENTNOR.   [Face page 422.

building. They form a uniform line facing south, the four eastern blocks being for women, and the six western ones beyond the chapel being for men. Owing to the slope of the ground, the ninth and tenth blocks have an additional storey, while preserving architectural uniformity ; they are also larger in other ways. Each block has a separate entrance behind, a covered verandah along the ground floor, and covered wooden balconies to all but the topmost storeys above. Patients occupy without exception a separate room on the south side ; the north side containing all the stair-cases, nurses' and servants' quarters. In the ground floor on the southern side are sitting rooms (some provided with billiard tables, bagatelle boards, etc.), mostly for the use of patients, a few of the rooms being reserved for nurses, and one for the chaplain and the hospital library. The eight smaller blocks consist of ground floor, first and second floors, and contain about twelve patients' bedrooms each. In the first block at the back is the superintendent's office ; but in other respects these blocks are almost identical.

No. 9 block is much larger, containing the administrative department, as well as eighteen patients' bedrooms. It is deeper as well as wider, and has, on the south side of the ground floor, a handsome dining hall, 70·×48 feet and 32 feet high, with parquet flooring, walls panelled for some distance and painted above, and handsome painted ceiling. It is provided with large windows, and lighted with incandescent gas mantles. At one end is a stage which is utilised for periodical concerts and dramatic entertainments. At the other is a large orchestrion, and portrait of the founder, Dr. Hassall. There are six long tables, two being for women. This and the chapel are the only parts where male and female patients meet. In the grounds there is a dividing line which they are not allowed to cross. On the north side of No. 9 block are consulting rooms, dispensary, board room, bath rooms, and quarters of the resident medical officers, as well as the kitchen depart-ment on the second floor. No. 10 block in many respects resembles the smaller blocks, although in a few details it is more modern. No. 11 block, which will probably be open next year, will have twenty-one bedrooms. Its floors and staircases are of teak, which will be " beeswaxed and turpentined," its walls of Parian cement with rounded angles, the doors and windows sunk flush with the walls, the balconies of iron and wide enough to permit of the beds being wheeled out on to them. Throughout the blocks which are open the walls are painted, the floors of deal with painted borders and linoleum in the centre except in the dining saloon. In the bedrooms the windows are alternately bow and flat ; the top floor has mansard windows. The dimensions vary from 1200 to 1400 cubic feet in the older to 1600 to 1800 in the newer blocks. The ventilation is partly by means of open windows, partly by the introduction of warmed air from the corridors through an aperture above the door. There are in addition ventilating shafts which are collected under the roof and exhaust the impure air, steam pipes being placed in the main shafts. By these means 5000 cubic feet of air per head per hour are admitted, at an uniform temperature of 62° F. The heating is by means of steam pipes, with radiators in the corridors and bedrooms. For lighting, gas is laid on from the town supply, but is not used in the bedrooms, where each patient has his own candle, which he lights at a jet in the corridor. The latter is lighted by paraffin lamps with reflectors. The furniture in the bedrooms is simple : white painted washstands, dressing tables, and chairs, and wire wove spring iron bedsteads. Excepting for food, coals and the like, there are no lifts ; but as far as possible the more robust patients are placed in the upper storeys. The total accommodation in the existing buildings is for 134, eighty-three men and fifty-one women ; but this includes two north rooms, but seldom occupied by patients. With the new block there will be room for 155. On the other side of the approach at the back, sunk below the road level, is the engine house for heating and pumping water. There is a large softening tank which has recently been added ; and it is proposed to make a reservoir on the hillside at a higher level. The water supply comes from a private well.

Like other British chest hospitals, the Ventnor Hospital nominally receives cases of bronchitis as well as phthisis, but practically the patients are all consumptives in an early or remediable stage. The irksome rules in force in most of the general hospitals are not adopted at Ventnor, patients rising at 8 o'clock and going to bed at 9 or 9·30 according to the season. The *meals* are four in number, at 8·30, 1, 5 and 8. The distance which the female patients have to walk is inconvenient in the case of feeble subjects. Those who are too weak to come to the dining saloon have some (or all) of their meals in the sitting room of their own block. At one time all used to dine in their own block ; but this system proved to be wasteful. *Sputa* are received out of doors into special Turkey-red handkerchiefs supplied by the institution, which are collected daily and boiled ; indoors, spitcups are used which contain a little water ; these are emptied into the drain without special disinfection, but the adoption of an incinerator is under consideration.

As regards open-air treatment, the Ventnor Hospital stands in a position intermediate between the ordinary chest hospital in a town and the true sanatorium. The ventilation, as already seen, is very abundant ; while the cleanness of the paint testifies to the purity of the air. The patients also spend much of their time out of doors, but no attempt is made to regulate the way in which they pass their time ; they walk or sit as they please, and probably spend too much time in the sitting rooms. Moreover, there is a rule whereby the windows are shut towards evening, and on a beautiful, warm October day I found them tightly closed in the bedrooms at tea time. Recumbency in the open air has not been systematically prescribed up to the present ; but when the new block is finished it will be adopted in suitable cases.

There is no systematic hydrotherapy. Cod-liver oil is largely prescribed, as well as drugs of various kinds according to necessity. There are sixteen nurses and two resident medical officers. Dr. Sinclair Coghill is the physician, Drs. Robertson and Whitehead assistant physicians to the institution. There are also a surgeon, an analyst, and consulting and examining physicians and medical referees to the hospital in various parts of the kingdom.

During 1898, 717 patients completed their term of treatment out of 848 received. Of these 78·2 per cent. improved, 11·3 per cent. remained *in statu quo*, 10·5 per cent. became worse or died. The average duration of stay was 62·5 days.

THE NATIONAL SANATORIUM FOR CONSUMPTION AND DISEASES OF THE CHEST,

at Bournemouth, is the largest institution but one of the kind in the British Isles, coming next in size to the Royal National Hospital for Consumption and Diseases of the Chest at Ventnor. Like the latter, it does not admit cases of advanced disease, and, nominally at all events, does not exclude the sufferers from asthma and chronic bronchitis.

It is situated in a valley at the back of Bournemouth, close to the Mont Dore Hotel. the grounds of the two institutions touching one another. The sanatorium has fairly extensive grounds, which are partly laid out as kitchen garden, partly as flower garden and shrubbery, with a fine lawn to the south of the institution, beyond which is a rapid descent to the bottom of the valley. The place is extremely well sheltered, not to say shut in, both by the neighbouring hillsides and by large trees, and shares in the climate of the less bracing parts of Bournemouth. The building consists of a long, low, stone building with ground floor and first floor, the former for men and the latter for women. The aspect is south-south-west, the entrance on the northerly side. At the eastern end is a large chapel attached to the institution. Behind the building, near the centre, is a yard, with various domestic offices shut off from the garden, and ending in a small mortuary and *post-mortem* room. Running from end to end of the building is a corridor; the patients' rooms are to the south of this, including on each floor a large dining saloon and day room, a ward for five patients, eight with three apiece and two single-bedded. The dining saloons are both cheerful and light, with a good view of the grounds; and the same may be said of the day rooms. The furniture is plain and simple, but suitable; walls painted, floor covered with linoleum, windows large and (in the upper rooms) some of them double for protection against rough weather. The wards and bedrooms are of good size and sufficiently lofty. The corridors are used after dark and in bad weather as promenades. The lower one is dark and not perfectly ventilated, as the chapel blocks the eastern end, while the western end has been blocked by a screen and converted into a smoking room for the men. It is lit by gas; during the daytime a large portion is darkened by the administrative block and staircase. The upper corridor is also somewhat darkened in the same way, but is free from obstruction at the western end. The lavatories and water-closets for the patients are dark and placed in most unsuitable positions, where it is almost impossible to adequately light and ventilate them. There was no great anxiety noticeable in keeping windows open at the time of my visit; and beyond the fact that the patients are encouraged to stay in the grounds as much as possible, no attempt has been made until recently to carry out systematic fresh-air treatment. A shelter has recently been placed in a warm corner of the grounds, a stone's throw from a large hedge of evergreens and much surrounded by trees and shrubs. This shelter, which is intended for the use of the women, has been substantially built of wood, with an overhanging roof, an opening in front and a door at the western end, the eastern end being closed. Owing to the lack of ventilation under the roof, it is sure to feel hot and oppressive in all but cold weather; but when considerably altered will be a useful adjunct to treatment. There is a fine terrace along the south side of the building, which, I am informed, is a good deal frequented by patients. With a few light and adjustable folding shelters this might easily be utilised for rest in the open air. The building is heated by open fireplaces and hot water pipes, the water supply and sewerage being that of the town of Bournemouth, part of which surrounds it.

The institution is closed during July and August, when in such a situation the heat would be almost unbearable to many people. There is a resident medical officer, and a nursing staff consisting of a matron and three nurses. The physicians are Dr. Snow, Dr. Frazer and Dr. Davison. There are also a consulting physician, consulting surgeon, surgeon and dentist attached to the institution.

The sanatorium is much used by various provident and friendly societies. There is room for sixty-two patients, an equal number of each sex. Admission is by a governor's nomination (which is available for twelve weeks, and can be renewed if advisable), and payment of 7s. 6d. weekly. During 1896, 162 men and 120 women were admitted.

THE CROMER EXPERIMENTAL SANATORIUM

was started by Dr. F. W. Burton-Fanning, of Norwich, in spring, 1895, at the Fletcher Convalescent Home, which is connected with the Norfolk and Norwich Hospital. It is a small home standing in three acres of ground, with beds for thirty convalescents, usually closed during winter and spring, while at other seasons its visitors are admitted for a three weeks' stay. The climate of this part of the coast is well known for its bracing qualities, which render it unsuitable to some who have a feeble circulation, but extremely invigorating to others. Cromer stands at the junction of the northern and eastern coasts of Norfolk, on chalk soil covered with superficial sandy and other deposits, and, like the rest of Norfolk, has a very small rainfall (about twenty-four inches). The town is about seventy

feet above the sea-level, but the home is about 250 feet
above the sea, and a quarter of a mile from it, while the
surrounding woods and hills shelter it from wind, so that
the place is both warmer and drier than some other parts
of the east coast.

Dr. Burton-Fanning wished to try the suitability of the
English climate for the open-air treatment of consumptives,
and made arrangements for six beds to be devoted to this
purpose throughout the year. The home has a sheltered
verandah and summer house, and Dr. Burton-Fanning had
another shelter erected in the grounds, with two movable
side walls of wood and glass. These were placed in position
every morning after observing the direction of the wind;
and patients then spent the greater part of the day in one
or other of the shelters, resting on long chairs. They
usually went out at 8·30 A.M. and came in at sunset; but
on particularly dry nights, both in winter and summer, they
remained out until 10 P.M. The results were most satisfactory. Out of twenty-four patients so treated, two were
apparently cured; relative cure in four more; great improvement in four; and only four died or failed to improve;
and in three of these there was temporary improvement.
Only one of the febrile cases (twenty-one in number) failed
to lose the fever, after from three to thirty-five weeks of
treatment, and all gained in weight. Several noticed that
they caught cold less frequently, and in no instance was
any bad consequence traceable. In all but one case the
diagnosis was verified by discovery of tubercle bacilli in
the sputum; and several of the cases came under treatment in too late a stage to hope for a complete restoration
to health.[1]

The Mundesley Sanatorium for paying patients, which is
to be erected under the auspices of Dr. Burton-Fanning, is
described at p. 346.

[1] See Lancet, March 5, 12, 26, 1898; Practitioner, June, 1898.

THE DOWNHAM SANATORIUM FOR THE POOR

consists of a small farmhouse where five or six patients can be accommodated, which was started some five or six years ago by Dr. Jane Walker, who has defrayed the expenses out of her own pocket. It is managed by Mrs. Beston, "to whose devotion and enthusiasm the success of the whole venture is largely due".

The cottage is separated by a small garden from the main road, and is an ordinary unpretentious but dry and well-built cottage ; the floors are boarded, the walls painted with a kind of silicate paint, the furniture and decorations of the simplest kind, the windows fairly large and kept constantly open. The garden is abundantly supplied with sheds facing in different directions, which can be used in wet or windy weather. The same *régime* as to diet and constant fresh air is followed as at the establishment for patients of the better class (see p. 344). The terms are 10s. to 15s. per week. Notwithstanding the extremely simple arrangements, Dr. Walker has had very good results, some cases greatly improving which had failed to do so in some other parts of England which are considered favourable to the recovery of consumptives. Out of seventeen patients treated since 1892, four died, four are under treatment, and nine have returned to their work and are doing well.

THE VICTORIA HOSPITAL FOR CONSUMPTION,

at Craigleith, near Edinburgh, which is the only hospital in the United Kingdom for the gratuitous treatment of consumptives strictly on the open-air principle, was opened in 1894, in connection with a free dispensary in Edinburgh itself under the care of Dr. R. W. Philip. It was originally a private mansion standing in 7½ acres of ground about a mile to the north-west of Edinburgh. It is placed on a gentle slope facing south, and is sheltered by a splendid belt of trees. The beautiful grounds are laid out with fine lawns and winding paths, and afford shelter in one part or another from every possible wind. There is at present accommodation for fifteen patients, seven male in two wards on the ground floor, and eight female in three wards on the first floor. The rooms have a cubic capacity of about 1000 cubic feet per head, some having nearly double this amount. Each has at least one large window which is open day and night, the larger ones having three. Heating is by open fires, the temperature being kept at as near as possible 60° F. The floors are plain and polished, the walls distempered. The furniture is of the simplest. The bedsteads have an open spring mattress, covered with a horsehair mattress. The bed tables are of glass and metal, with open shelves, one of which is protected by a thin strip of cotton.

The treatment is according to hygienic rules. Every patient spends many hours out of doors every day. Those who are febrile or have much circulatory disturbance are carried out to reclining chairs, or on their beds. In other cases graduated exercise is adopted, walking, gentle cycling, mild golfing, quoit throwing, dumbbells, Indian clubs, or gentle breathing exercises, according to the condition of each, and the effect on the pulse. Patients rest before and after every important meal. Every patient is either sponged or bathed in more or less cold water. Occasionally systematic massage is applied. The clothing is simplified as far as possible, Shetland woollen materials being much used. Five *meals* a day are given, two consisting of soup or warm milk. Stimulants are given to pyrexial cases as a rule, and to those with much disturbed circulation. *Sputa* are received solely into spitcups containing 1 in 20 solution of carbolic acid, or into a modification of Dettweiler's flasks. These are cleansed with boiling water and poured down the water-closets. Handkerchiefs are not used for sputa, but occasionally rags are used for wiping which can afterwards be burnt. Of drugs, cod-liver oil and arsenic are the most used.

Dr. Philip reports that he has seen nothing but good from the treatment, even in cases with considerable pleural effusion. The climate is that of Edinburgh, temperate, but changeable, with a fair amount of mist and wind; but the weather is seldom allowed to keep suitable cases indoors· The main results have been an improved colour, better appetite, disappearance of night sweats, diminution of fever, increase of body weight and diminution in cough. The patients at present are seldom able to stay more than two months; but Dr. Philip considers from four to six months desirable.[1]

THE BRIDGE OF WEIR HOSPITAL FOR CONSUMPTION,

which is being built by Mr. Quarrier in Renfrewshire near his Orphan Homes, is to accommodate from thirty-five to fifty patients in a number of wards, ten of which will be single-bedded. The walls are to be of hard and smooth plaster, with rounded angles, the floors to be waxed. When funds permit, eight or ten similar pavilions will be erected. Mr. Quarrier's scheme includes the erection of a dispensary in Glasgow.[2]

THE NATIONAL HOSPITAL FOR CONSUMPTION FOR IRELAND,

at Newcastle, in the Wicklow Hills, is situated three miles from the sea, on the southern slope of a hill, which shelters it from the north and to a less extent from the east. The hospital is 270 feet above the sea-level; to the south-west and west, at a distance of a mile or two, is a chain of hills 700 or 800 feet high. The soil is of gravel, in the deeper parts of which are many springs of water. The climate is mild, humid, and equable, with from 77 to 84 per cent. mean daily humidity, about 40 inches rainfall, and 195 rainy days. The prevailing winds are north-west and south-west.

The hospital consists of an administrative block with two

[1] See *Brit. Med. Journal*, 23rd July, 1898.
[2] See *Brit. Med. Journal*, 15th Sept. and 1st Dec., 1894; *Lancet*, 12th Dec., 1896.

blocks for patients, each of the latter containing twelve beds. Provision has been made for the future erection of six more blocks. The sewage passes into a closed cesspool, the overflow from which passes into a branching drain deep in the soil, where it percolates in every direction.

Treatment is by open-air methods, with abundant food, attention to the functions of the skin, moderate exercise, and sufficient rest in bed.

The physicians are Dr. O'Carroll, Dr. Parsons, and Dr. Coleman ; the resident physician, Dr. Steede.

Out of 100 consecutive cases in which tubercle bacilli were identified in the sputa, thirteen are stated to have been very much improved, with a probability in many instances of permanent arrest ; twenty-eight were much improved, with a probability in some instances of permanent arrest ; thirty-six were improved. In all these three classes the weight increased, the general condition improved, and the disease was more or less checked. Fifteen remained in much the same condition, although some became somewhat stronger. Six became worse, and two died. These details are from a paper by Dr. Steede at the Dublin Congress of the Royal Institute of Public Health in August, 1898.

THE FORSTER GREEN HOSPITAL,

for consumption and diseases of the chest, was opened on 30th October, 1897, at Fortbreda, on the Castlereagh Hills, a short distance from Belfast, but sufficiently far removed to ensure a pure atmosphere and a healthy situation. Erected through the munificence of Mr. Forster Green, who contributed over £13,000, it has been amalgamated with the Belfast Hospital for Consumption and Chest Diseases, which has existed in the city since 1880.

The soil is pervious, the situation elevated, airy, yet fairly sheltered, with a southerly aspect. The building at present has accommodation for forty patients, but has been arranged with a view to further extensions. It consists of a ground floor and three upper floors. On the ground floor is a dining saloon, 30 × 16 ft. ; a kitchen, 22 × 17 ft., with scullery, pantry, etc. ; a dining room for the staff, 24 × 12 ft. ; a smoke room and store room. On the first floor on one side is a board room, 17 × 13 ft., and on the other four wards, each 23 × 17 ft. On the next floor are sitting and bed rooms for the nursing staff and matron, together with seven wards. The top floor contains bath rooms and servants' bedrooms.

The building is lighted with electricity, and ventilated according to the *plenum* system, a minimum of 5000 cubic feet per head per hour being pumped into the building. The air can be delivered at will, either all heated or with various proportions unheated. There is a good lift. The water supply is pumped up by the electric motor ; and all the appointments are stated to be according to the most modern standards.[1]

[1] *Brit. Med. Journal*, 6th Nov., 1897 ; *Lancet*, 13th Nov., 1897.

CHAPTER XLVIII.

THERE is scarcely any suitable provision in this country for middle-class consumptives. Until quite recently there were no sanatoria for such patients; and even now there are only three, with a total accommodation for twenty patients, and three others which are being built. Another sanatorium was projected by Dr. Plater Long at Crowborough in Sussex, but I have been unable to obtain recent information about it. I have myself been preparing during the last two years, with the help of architects, engineers, and other experts, the plans and estimates for a sanatorium for the less wealthy middle classes. A most suitable site has been found, and the sanatorium would have been started some time ago, but for the difficulty in satisfying the demands of probable supporters without offending against medical etiquette. The ordinary layman cannot understand the objections to medical ownership of such institutions; nor is it easy to raise a considerable sum of money without publishing a list of medical supporters or obtaining written promises of support, both of which would be objected to by self-respecting physicians. However, these difficulties are in a fair way to be overcome in the near future, and I hope that the sanatorium will be started before long.

The following is a list of the existing sanatoria for the middle classes, together with those in process of erection :—

(335)

BRITISH SANATORIA FOR PAYING CONSUMPTIVES.

			Feet.	Beds.
Bournemouth	Pool Road Sana-torium	Dr. Fras. Pott	120	8
,,	Sunny Mount Sanatorium	Dr. Johns	150	4
Denver, Norfolk	Denver Sanato-rium	Dr. Jane Walker	55	8
Cotteswold Hills	Cotteswold Sana-torium (being built)	Dr. Pruen and Mr: Hartnell	800	10
Mundesley, Nor-folk	Mundesley Sana-torium (being built)	Dr. Burton-Fan-ning and Mr. W. J. Fanning	200	25
Ringwood, Hants	Ringwood Sana-torium (being built)	Dr. Mander Smyth	?	?

THE BOURNEMOUTH SANATORIA.

There are two private sanatoria at Bournemouth, belonging respectively to Dr. Pott and Dr. Johns, and a third is being built by Dr. Smyth at Ringwood in the New Forest, the " hinterland of Bournemouth ". There is also an hotel [1] which with its grounds has been specially adapted for the open-air treatment: a home for invalid ladies of small means; besides other institutions for the poor.

Bournemouth lies on gravel resting on a subsoil of sand, at least fifty feet thick, which in turn rests on the chalk. Its climate is warm and equable, and in hot weather somewhat relaxing. It is abundantly sheltered from wind by various hills and pine woods; and although containing a population of 40,000 inhabitants is less "towny" than many other places of the same size, owing to the large area over which it is scattered. It enjoys a large proportion of sunshine, and a lower humidity than most of the health resorts of the south of England. There is a complete system of drainage, and a good water supply.

[1] Branksome Towers Hotel.

Dr. Pott's Sanatorium,

on the Poole Road, Bournemouth, was formed out of a pair of semi-detached houses, which were thrown into one and altered for the reception of consumptives. Dr. and Mrs. Pott live in one half, while the other is occupied by patients under the care of an experienced matron. Those who know Bournemouth need scarcely be reminded that a large part of the town consists of detached and semi-detached villas of good size with large gardens. It is in an open residential district of this kind that Dr. Pott's sanatorium is placed, the altitude being about 120 feet above the sea-level and the ground fairly level. The garden stretches southwards from the house towards the main road, so that less dust is blown up to the sanatorium. In the garden two or three simple wooden shelters have been erected, which can be taken to pieces and stored away.

The building is in the shape of a hollow square with the north side omitted. It has a long south front, seventy feet long, with short wings diverging from the two corners. Between these a verandah has been built out, which is twelve feet deep, with two large glass windows in the roof. In front of this a wire netting has been stretched, covered with large strips of transparent canvas to increase the privacy, and ward off the stronger southerly gales while permitting free access of fresh air. Comfortable cushioned couches with wire bottoms are placed in the verandah for rest in the open air. Over the verandah is a balcony, partially roofed over. The patients' bedrooms occupy the south side, on the ground floor and first floor. Behind them are the corridors, of which the lower forms a fine hall from end to end of the house, with windows at each end. This is furnished with chairs and other furniture, so that it can be used for rest or exercise by the patients when they cannot be out of doors. The main staircase leads off one end of this corridor. The dining room is on the side of the house to the north of the corridor. There are no other sitting rooms for patients,

22

who are expected to spend most of their time out of doors.

The bedrooms are all of good size, the smallest being 14 × 12 and 12 ft. high, with a capacity of over 2000 cubic feet, and some being considerably larger. The floors are painted along the border, with linoleum in the middle ; the walls painted with *duresco* (a silicate paint); the windows next the balcony and verandah are large " French " windows down to the ground ; in other rooms they are of the ordinary English pattern, but provided with special catches. There is an open fireplace in every room. Mouldings and ledges have been for the most part removed. Carpets and rugs are only sparingly used. In the lower corridor, cork matting in squares is employed, presenting a washable warm surface, while it prevents noise. The furniture is all arranged with a view to proper cleansing, standing away from walls, and being raised above the ground, or else easily movable, dead space and awkward ornamentation being avoided. In the bedrooms the furniture is mostly of white painted wood, the bedsteads being of iron. These are placed away from the wall, with the head towards the window, a tall screen being placed round the head of the bed.

The heating is by means of open fires ; the lighting by gas, paraffin lamps and candles. No special ventilating contrivances are found necessary. The water-closets, lavatories and bath rooms are all next outside walls on the north side. The staff are lodged partly in the northern wings, partly under the roof. There is accommodation for eight patients; a few rooms of smaller size, and not on the south side, are available for the use of their friends.

The rooms are rubbed down daily with a damp cloth. The *sputa* are received into red hour-glass shaped vases containing weak carbolic solution, with plain glass covers fastened on with gummed paper. To be used this must be torn, so that those which need cleansing can be recognised. They are collected daily, and treated with strong carbolic solution, the sputa being then poured down the water-closets. The patients also use Japanese handkerchiefs, which are afterwards destroyed.

None but consumptives are received, and no hopeless or seriously affected cases. They are treated continuously with fresh air, in and out of the house ; spending seven or eight hours out of doors on an average. Hydrotherapy is not much used, excepting in the form of hot or cold baths. Four *meals* a day are provided, including two of three or four courses each. Extra milk is also given on waking. Patients take as a rule about three pints of milk per diem, all of it being sterilised by the dairy which supplies it.

The sanatorium is a few minutes' walk from West Bournemouth Station. The usual charges are five guineas per week for board and lodging, and two guineas for medical attendance ; drugs, wine and washing being extra.

Dr. Pott, while superintending the patients in his sana-torium, is also engaged in a large general practice. Not-withstanding this drawback, he has had a very gratifying measure of success.

DR. JOHNS' SANATORIUM,

at Sunny Mount, Meyrick Park, Bournemouth, was started in 1897, largely under the advice and guidance of Dr. Arthur Ransome. Meyrick Park occupies a wide winding valley, which runs inland to the extreme limit of Bournemouth, and includes the golf links on one side, the football, cricket, and tennis grounds, and the bowling green, on the other. Crossing the end is the railway line, immediately beyond which, on a small hill commanding a view of the whole valley, stands Dr. Johns' sanatorium. Behind it are the pine woods of Winton, some 500 acres in extent ; hills rise to the east and west, while the town of Bournemouth lies at the end of the open valley to the south. The house stands in over an acre of ground, in which several shelters have been erected. There are a few villa residences near by, in

one of which Dr. Johns lives. The elevation is about
150 feet above the sea-level. The main road in front
of the house is nearly level; in other directions it is
hilly.

The sanatorium has a southerly aspect, and consists of
a ground floor, first and second floors, the latter being
under the roof. At the eastern end of the south front
the large semicircular window front of the drawing
room forms a conspicuous object, and next to it the
glass verandah which leads into the central hall. The
latter is 40 ft. long and 15 ft. wide, and reaches to the
back of the house. On the other side of the verandah
is the dining room, which, with the kitchen department
behind, forms a separate western half, so that the smell
of food may be easily prevented from reaching the rest
of the house. The staircase is at the back of the central
corridor, where also is the main entrance. Two large
windows on half landings with window seats help to
break the ascent to the bedrooms. These are on the first
and second floors, the one above the drawing room having
a large covered balcony, and that on the top floor a
somewhat smaller one.

There is accommodation for four patients, as well as
a few friends; the smallest patients' bedroom having over
2000 cubic feet air space. The floors are covered with
linoleum, or with cork carpet; in some parts they are
plain painted. The walls are of washable paper. The
furniture is not very different from that in an ordinary
middle-class household. The heating is by open fireplaces,
the lighting by gas, paraffin lamps, and candles. There
are no special ventilating contrivances.

For the *sputa* spitcups are used containing strong solution of chinosol. The contents are sometimes sent to the
destructor (which is not far distant), sometimes burned
in the house or poured down the water-closets. Floors and
furniture are cleansed with cloths dipped in one per cent.
solution of chlorinated lime. Walls on the patients'

departure are rubbed down with bread, which is burned, the room being also fumigated with sulphur.

The sanatorium is managed by Dr. Johns' sister, who is a certificated nurse. The charges are from 4 to 7 guineas, medicines, beverages and personal washing being extras. Friends and relatives accompanying the patient pay 3 to 4 guineas ; but with so small an accommodation their presence is undesirable. The sanatorium is about twenty minutes' drive from East Bournemouth Station.

THE COTTESWOLD SANATORIUM FOR CONSUMPTION,

on the Cotteswold Hills, near Cheltenham, has a southerly aspect overlooking the Stroud valley. It is situated within 200 yards of the line separating the Severn and Thames valleys. From this line a view of the entire Severn valley is obtained, stretching from the Shropshire hills in the north to the Bristol Channel on the south-west. The entire range of the Malvern Hills is seen in the north-west, and the Welsh hills beyond them. The descent to the valley of the Severn is very steep, but the walks to the east and west are gentle slopes upwards ; that to the Thames valley on the south a gentle slope downwards. The soil is oolitic limestone, from 150 to 300 feet thick. The climate of this part of England is bracing, although the neighbouring health resort, Cheltenham, can scarcely be called so.

The sanatorium stands 800 feet above the sea-level, in its own private grounds nearly seven acres in extent, well wooded with fir and beach. The grounds stand nearly in the centre of a tract of common land about 1000 acres in extent, which is similarly wooded. Protection against cold winds is afforded by the trees, as well as by the rise of the ground fifty feet behind the sanatorium to the north and east. The nearest public road (a branch road) lies behind a small hill 100 yards away to the north ; and for some distance round the sanatorium there is nothing but trees, grass and moss. Any trees which impede the view

to the south and west will be cut down, so that the site
will be sunny as well as sheltered.

The buildings consist of residential blocks for the patients,
an administrative block, servants' block and engine house
block.   There is at present one residential block, and the
second is expected to be ready in about two months' time.
If the sanatorium answers the expectations of its founders,
others will be added from time to time, until there is room
for about fifty patients.   All the buildings are on one floor,
each residential block being of wood on a stone foundation,
with rooms for five patients.   The walls are boarded
outside and in, with felt between.   The inside walls,
floors and ceilings are of varnished pine, the walls and
ceilings of the bedrooms being covered with canvas over-
laid with varnished washable paper.   The bedrooms are all
on the south-west side; behind them runs a corridor, and
behind this is a sitting room, short passage, bath room,
store room for linen and other things, and nurse's room.
The well-lit and ventilated attic contains hot and cold
water tanks.   In front of the block runs a glass covered
verandah, 9 feet 6 inches wide, and to the sides and back
a narrower one of the same kind.   The ends of these
verandahs are protected by screens of wood and glass.
Each bedroom is 12 × 8 feet and 10 feet high (= 960 cubic
feet).   Windows are $5\frac{1}{2}$ × $3\frac{1}{2}$ feet, starting $3\frac{1}{2}$ feet from
the ground, and consisting of French windows with a
hinged upper transverse pane.   Outside are louvre shutters.
The furniture is plain and simple ; one rug by each bedside.
The corridor has a door and window at each end ; the far
end opens by a trapdoor into the attic ; one end is blocked
by a projection from the nurse's room.   The sitting room
is $15\frac{1}{2}$ × 8 feet.   Behind it in the back corridor is the earth
closet.   Heating is by hot water pipes, which pass through
the stone foundations under the floor, with radiators in
every room, corridor and earth closet.   Lighting is by
electricity, the dynamo being worked by an oil engine.
Ventilation by open doors and windows.   The earth closets

are of Dr. Vivian Poore's pattern, but the pails will be emptied once or twice daily. There are electric bells in every room to the servants, the matron, house physician and male attendant; similar bells in the verandahs and earth closets. The bath room is supplied with hot and cold water and douche; in the second block there will be a hot and cold douche in each bedroom.

The dining block contains a dining room, kitchen, scullery, linen room, pantry, larder and servants' enclosed corridor. A verandah runs in front of it, which continues the line of the posterior corridor of each residential block. The servants' block also has quarters for matron and the house physician. A male attendant sleeps in the engine house block.

The water supply comes from two springs in the private grounds of the lord of the manor, who is also lessor of the sanatorium ground. These supply ten gallons a minute, which are pumped up to the sanatorium 160 feet higher up by an electromotor. The quality of the water is stated to be excellent, and not too hard.

Sputa will be received into Dettweiler's flasks and spittoons containing disinfectant; the contents burnt in the furnace and the receivers boiled. Linen to be disinfected by boiling. Rooms by washing and re-varnishing or re-papering.

Treatment as at Nordrach; windows constantly open, except when dressing or undressing. Patients will stay outside all day long, and take as much exercise as they are fit for, returning downhill. Dr. Pruen states that there is always shelter from driving rain in one direction or another. Patients are not encouraged to congregate in common rooms. Three rather heavy *meals* per day are provided. Hydrotherapy is only adopted in the shape of daily baths, unless the patient is too ill, when his body is rubbed with spirit and salt. Attendants and friends are only exceptionally permitted to live at the sanatorium.

The staff consist of a resident house physician, a resident

matron, and extra nurses according to need. Servants, as
already stated, in separate block. Dr. S. T. Pruen and Mr.
C. Braine Hartnell, M.R.C.S., come over daily from Cheltenham to visit the patients.

The charges are 4½ guineas weekly, extras being beverages
and personal washing. The sanatorium is seven miles by
road from Cheltenham, Gloucester, or Stroud. It is on the
telephone system.

THE DOWNHAM SANATORIA.

Two small sanatoria have been brought into existence at
Downham, in Norfolk, under the auspices of Miss Jane
Walker, M.D. One of these, which is intended for poor
patients, is described at p. 330. The other, for those able
to afford a small inclusive fee, is situated about one and
three-quarter miles from Downham Market, twenty minutes'
drive from Downham Station on the Great Eastern Railway.
It is on the lower greensand, on sloping ground about fifty-
five feet above the sea-level, and consists of a substantially
built farmhouse in a large garden. The surrounding country,
which is very flat, is mostly grass land. There are some
pine woods not far off and a large extent of common land.
The population is scanty and the cottages scattered. The
rainfall is low (20 to 24 in.), but there is very little dust.
Shelter against wind is obtained by means of high hedges
protecting from the east and north. Distant shelter is
defective. A well-built outdoor shelter has been placed in
the garden, facing south. It is 25 × 12 ft., constructed
of pitch pine, with a glazed roof.

The house is spacious and airy, and at all times very
fresh and dry. There is a large common room on the
ground floor, facing west and north, which is used as a
dining room. Close communication between the patients
is discouraged except out of doors, so that no other common
room is provided excepting for visitors and the use of the
managers. There is no basement, the kitchen being on the
ground floor. The average size of the bedrooms is 17 × 14

× 9 ft. They have various aspects on the first and second floors. The walls are distempered, which is freshly done before each patient's arrival. The floors are stained and cleaned with beeswax and turpentine, and also frequently washed. There is a square of linoleum in some of the rooms, and strips of carpet in the bedrooms. The furniture is of the simplest description. The rooms are ventilated by sash windows and (in most cases) by chimneys. The former are kept wide open both top and bottom night and day. In one of the rooms a double layer of "greenhouse shading" stretched over a light wooden frame has been substituted for a more substantial door. This is sufficiently opaque to ensure privacy, while allowing of perpetual ventilation and periodic removal for cleansing. The heating is by open fires or (in a few of the rooms) by oil stoves. In the common room a closed smoke consuming stove for burning anthracite is used. The lighting is by lamps and candles. Rain water is used for washing purposes, and hard chalky spring water for drinking. The water-closets are on the pail system, and are out of doors.

Eight patients in all can be accommodated in the house; extra rooms are obtainable in neighbouring cottages. None but consumptives are admitted. Patients who expectorate are provided with sputum cups, which are scalded twice a day or oftener, all the sputum being burnt. In some cases metal cups with stiff paper linings are used. The linings in this case are burnt, the cups being scalded. Both table napkins and handkerchiefs are used. These, and other infected linen, are put into boiling water. The *meals* are three in the day: breakfast at 8, dinner at 1, supper at 7. The food given is of a substantial nature. The morning meal is a good English breakfast with a pint of milk. Dinner consists of two meat courses, pudding and cheese, and a pint of milk. Supper of one meat course and soup or fish with pudding and a pint of milk. Each patient is given an ample helping and expected to finish it.

The sanatorium is managed by Miss Wright and Miss

Preston, one of whom looks after the patients, and the other superintends the preparation of the food. The resident medical officer is Miss Hawker, M.B. Dr. Jane Walker pays a weekly visit, and is in constant communication with the sanatorium.

The charges are three to four guineas a week; drugs, beverages and personal washing being extras. Downham Station is 87 miles or 2½ hours by rail from London.

THE MUNDESLEY SANATORIUM

is to be erected half a mile from Mundesley Station on the Great Eastern Railway and seven miles from Cromer, on the coast of Norfolk. The soil of the district is sand resting on the chalk, while the whole district is known to be relatively dry and bracing. Twenty-five acres of land have been acquired, which form a strip almost entirely on the southern slope of a ridge which runs east and west, parallel to the coast and about half a mile from it. Gardens and walks are to be laid out, which will be sheltered not only by the high ground behind, but also by plantations. The site of the proposed building is at the foot of a knoll on which are fir trees, and which will afford additional protection from the north, while the curve of the ridge gives shelter from the east. From the summit of the ridge a good view is obtained over the sea and surrounding country, while to the south an extensive tract of more or less flat agricultural land is commanded. The sanatorium will be approached by two private roads, and there is no main road within a quarter of a mile of it, securing isolation and freedom from dust. The elevation is about 200 feet above the sea-level; the position freely open to the sun. It is believed that the situation combines the advantages of the bracing sea air of the east coast with efficient protection from the most prevalent winds.

The building is to accommodate fifteen patients, and will have large sitting rooms, verandahs and balconies, washable coverings for the walls and floors, with rounded angles and nothing calculated to harbour dust. The heating is to be by open fires and hot water pipes, the lighting by electricity; dry earth closets will probably be provided. There is water laid on from the public mains.

There will be one or two resident nurses, Mr. W. J. Fanning the resident medical officer. Dr. F. W. Burton-Fanning of Norwich will act as consulting physician to the establishment.

A separate weekly charge will be made for medical attendance, the other charges being sufficient to cover expenses and pay at the most 4½ per cent. to the subscribers. It is believed that from three to four guineas a week will satisfy these conditions.

Dr. Burton-Fanning's experimental sanatorium for poorer patients is referred to at p. 328.

CHAPTER XLIX.

BRITISH NEEDS AND BRITISH RESOURCES.

THERE has been a steady and considerable reduction in the mortality from consumption during the last sixty years in this country. In 1838 the death-rate from this disease was over 38 per 10,000 living, in 1896 it was only 13·05, or a diminution of nearly two-thirds. This has been attributed with good reason to the establishment of special hospitals for diseases of the chest, and to general sanitary improvements in house construction, drainage of the land, ventilation of workshops and factories, and the like. A vast amount of money has also been spent in the erection of convalescent homes, which no doubt have helped to stave off the decline of many a consumptive, although such patients are not welcome or desirable visitors to these institutions, nor indeed adequately provided for in them.

Great Britain was probably the first country in the world to establish special hospitals for the treatment of consumptives. Her first seaside sanatorium for scrofula was founded at Margate as early as 1791, and the Royal Hospital for Diseases of the Chest in London in 1814; while the Brompton Hospital for Consumption and Diseases of the Chest with its 321 beds is almost the largest of the kind in the world.

Six years ago, when Germany's first sanatorium for the poor was erected, England was unrivalled in her provision for the consumptive poor. There were at that time no such institutions in any other part of the continent, whereas, in the British Isles, there were some seventeen special hospitals and nursing homes in existence, with over 1100 beds, besides other institutions open to consumptives, though not exclusively devoted to them. Since that time, however, great strides have been made in Germany, so that she will very soon be far ahead of England both in the number and the character of her institutions of this kind.

BEDS FOR THE CONSUMPTIVE POOR.

	Urban.	Country.	Total.
Great Britain . .	671	599	1270
Germany . . .	—	nearly 1300 } 2400	nearly 1300 } 2400
„ (being built)	—	about 1100 }	about 1100 }

Germany has at the present time about twenty sanatoria in working order with an aggregate of nearly 1300 beds, while in the near future there will be nearly double the number of sanatoria with over 2400 beds, besides a number of open health colonies which already exist in the country under medical supervision, where consumptives are received for hygienic treatment. Moreover, the majority of these sanatoria are exclusively for consumptives, and in extremely well-chosen spots, combining purity of atmosphere with shelter against dust and wind ; whereas the British institutions are mostly urban, and shared with sufferers from bronchitis and heart disease. Even allowing for the difference of population between Germany and England, which is roughly as five to four, it is evident that in the absence of extensive schemes like those of Germany, our accommodation for the consumptive poor will very soon be far behind that of our continental neighbour. It is equally certain that we have not sufficient for our needs. Prof. v. Leyden estimates that if each consumptive has three months' treatment, there should be one bed for every 1000 inhabitants.[1] If this be correct for Great Britain and

[1] Quoted by v. Weissmayr, *Wiener Kl. Wochenschrift*, 14th Jan., 1897.

Ireland, we should have about 38,000 beds for consumptives of all classes, or perhaps half this number for those among the poor. It is possible that this is an over-estimate; but in any case our provision is much too small.

One serious consequence results from this lack of suitable accommodation. Owing to the large number of applicants for admission, patients are often kept waiting for weeks before they can be properly treated ; and the average duration of their stay is reduced to a few weeks—six weeks or two months—a period insufficient either to cure them or to instruct them in the hygienic rules on which their well-being depends. Moreover, the patient is often ineligible for re-admission during the same year. Contrast this with the system at the French hospitals for consumptive boys, where the patient remains until he has recovered; or with that in Germany, where the consumptive is promptly sent for three months' systematic treatment in quite an early stage, with the possibility of a further three months or more if necessary. Were our patients always sent on to model villages, or even for several months' treatment under adequate medical supervision at a well-situated convalescent home, this reflection would lose its sting. But under existing circumstances our treatment of consumptives can only be called halting and inefficient in comparison with that in Germany. Our consumption hospitals are splendid institutions, and have done most useful work ; but the majority are situated in an impure atmosphere, and by them-selves are utterly insufficient for the work they have to do.

The cure for this state of affairs is not the erection of new consumption hospitals in our towns, but the provision of country sanatoria, perhaps in connection with existing institutions, where the more hopeful cases can be sent on to complete their recovery ; and of many more nursing homes for the chronic and incurable cases. To let the hopeful cases return to their insanitary conditions of home life after a short period of treatment, is to deprive them of their chance of recovery; while to allow a chronic consumptive of the poorer classes who is unable to work to remain at home greatly increases the risk of infection, while it entails much misery on those who live with him.

The consumptive is an undesirable visitor at a convalescent home; but there would be no objection to the transformation of a number of these institutions into sanatoria under medical supervision. Every large town and district of Germany has its own sanatorium ready or in process of erection; and a similar arrangement should be made in this country.

Many relapses are occasioned by the return of consumptives to an unsuitable occupation or neighbourhood. To prevent this, model villages should be established under medical supervision. Many occupations known to be injurious to consumptives might be rendered harmless by altering the conditions under which they are performed. In order to prevent the soiling of materials by smoke and dust, trades and occupations are carried out behind closed windows which might equally well be done in clean houses with open windows in the country; and even short of the erection of model villages, much could be done (and, indeed, has been done) by persistent medical pressure to improve the conditions of life.

To ensure a reasonable amount of uniformity, it would be advisable to appoint a few medical men (preferably such as are acquainted with public health as well as with sanatorium treatment), each to supervise a district and regularly inspect the various sanatoria and model villages within it. Such supervisors would confer with the local medical officers and advise them while reporting to some central authority.

To prevent disobedience in hygienic matters on the part of convalescent patients, certain privileges might be continued after their leaving the sanatorium (as with members of benefit societies), which would be forfeited on breach of rules. As in the case of convalescent homes, the railway companies would, no doubt, allow special reductions in fare to those going to a sanatorium.

For the children of consumptive parents, it would be advisable to establish special *schools for delicate children,* where the laws of health were more scrupulously obeyed.

Tradesmen abroad frequently subsidise a sanatorium in order to secure a bed for themselves or their family in the event of their becoming consumptive. In several instances, too, large manufacturing firms in Germany have erected sanatoria as well as convalescent homes of their own for the benefit of their workmen.

Turning now to the arrangements for middle-class consumptives, we find that Germany is even further ahead of this country than as regards the poor. This is partly because of a feeling here that the middle-class patient can afford to pay for home treatment; partly because our climate is regarded in many quarters as unsuitable; partly also because the disadvantages and difficulties of home or hotel treatment are not sufficiently realised. Even where patients can afford to go abroad, there are often grave objections to their doing so. That our climates are suitable for the treatment of consumptives is sufficiently proved by the results already achieved under great difficulties in various parts of the country—Ventnor, Bournemouth, Norfolk and elsewhere—and by the long list of distinguished physicians who have publicly and privately advocated the hygienic treatment in our own country.[1]

Home treatment of consumptive patients during febrile or complicated stages is far more expensive than is usually realised. Allowing for the salary and keep of a trained nurse (who is needed during febrile stages for a part of both day and night, unless relatives can attend to the nursing), and for an average of two visits a day from the medical man, it will not be safe to reckon the cost of treatment during critical stages at less than £5 per week, ex-

---

[1] Among these may be mentioned Sir Richard Douglas Powell, Dr. Theodore Williams, Dr. Hermann Weber, Dr. Ransome, Dr. Kingston Fowler, Dr. J. A. Lindsay of Belfast, Dr. Philip of Edinburgh, and nearly every physician attached to the London chest hospitals.

clusive of rent and ordinary household expenses, or of the
cost of rooms in lodging-house or hotel, if the patient is
away from home. And as the physician in such cases does
not live under the same roof, he is quite unable to exercise
efficient control over the hygienic details which so often
make the difference between recovery or the reverse, and in
which consists the special educational value of a sanatorium.
Patients who pay from £4 to £6 per week, as at most of
the continental sanatoria, have a right to expect not only
proper sanitary conditions, good medical attention available
at any moment, a good and suitable dietary and efficient
nursing, but also an atmosphere of unexceptionable purity,
free from boisterous winds, and grounds which are of such
a size and character as to allow of graduated walking exercise, as well as rest in the open air.   At the present moment
I know of no sanatorium in England which combines all
these advantages without using public roads, although it
is to be hoped that some of those in course of erection will
do so.

There is still greater need of good sanatoria which will
offer these advantages at a moderate charge. There are
many patients not properly admissible to institutions for
the poor, who yet cannot afford to pay more than an inclusive charge of from two to four guineas per week. The
poor curate, the briefless barrister, the struggling doctor,
governess, teacher, actor, artist, the banker's or mercantile
clerk is utterly unable to afford more than a very modest
fee, and has frequently been driven through lack of other
resources to one of the consumption hospitals which are
neither intended for such a class nor able to provide sanatorium treatment. There is in fact no adequate provision
for this class of patient, who is often worse off than the
artisan or domestic servant. Such provision at low cost
can only be made by a sanatorium with accommodation for
a dozen or more patients; for in many respects it costs as
much to provide for two or three patients as for twenty or
thirty. The administrative department, the efficient cook,
23

the housekeeper, the resident medical officer, the means of shelter against dust and wind and rain have to be provided for one patient just as much as for a dozen, and it is only by dealing with a number that economy can be effected, variety introduced, and dulness prevented. A glance at the following table will show the inferiority of our provision for the middle classes as compared with that of Germany :—

SANATORIA FOR PAYING PATIENTS.

	Sanatoria.	Beds.
Germany	17	Nearly 1200
England	$\left.\begin{matrix} 3 \\ 2 \end{matrix}\right\}5$	$\left.\begin{matrix} 20 \\ 35 \end{matrix}\right\}55$
,, being built .		

Only twelve beds are at present available in sanatoria for middle-class consumptives at less than five guineas per week. To satisfy the needs of our present population, even for a short course of treatment, there should be several hundred. By combination and co-operation alone can such a deficiency be supplied; for the task is beyond the power of unaided individuals to accomplish.

The establishment of various societies for the suppression of tuberculosis, one of them under most distinguished patronage, is but one of many signs that Great Britain is at last waking from her satisfied slumbers, and preparing to again take her place in the van of the nations. Our country's sanitary past has been great and fruitful; and there is every reason to hope that with growing consciousness of the possibility of destroying this dread scourge of humanity, by the abolition of town smoke, the improvement of our dwellings, the better ventilation of rooms and streets, the admission of sunshine into our midst, the inculcation of more rational habits of life, the destruction of sputa, the erection of sanatoria, and in many other ways, she will gradually prepare for herself a still more great and glorious future.

BIBLIOGRAPHY.

ADIRONDACK COTTAGE SANITARIUM, The, *New York Tribune*, 17th June, 1894.
Annual Reports of various Sanatoria—mostly omitted.
ARMAINGAUD.—Prophylaxie de la Tuberculose. Paris Congress, 1892.
—— Ligue Préventive contre la Tuberculose. Paris Congress, 1893.
ASCHER, B.—Die Lungenheilstätten und die Invaliditäts-Versicherungsanstatten, *Deutsche Medicin. Wochenschr.*, 1895, No. 19.
AUFRECHT.—Zur Verhütung und Heilung. der Chronischen Lungentuberculose. Vienna, 1898.
AZIÈRES.—Sur la Création de Sanatoria pour les Phtisiques Indigents, *Rev. d'Hyg.*, Paris, 1898, xx., and *Gaz. des Eaux*, Paris, 1898, xli. 171.

BARATIER.—La Tuberculose au Village, *Tribune Médicale*, 1897, 2 S.
BARTH.—Therapeutique de la Tuberculose. Paris, 1896.
BEALE, E. Clifford.—*See* Harris, v. Jaruntowsky.
BEAULAVON.—Contrib. à l'Étude du Traitement de la Tuberculose Pulmonaire dans les Sanatoria. Paris, Bataille et Cie.
—— Les Sanatoria pour Phtisiques Indigents a l'Etranger, *Revue de la Tuberculose*, Dec., 1896, and April, 1897.
—— Nos Sanatoria Français pour Tuberculeux, *France Méd.*, Paris, 1898, xlv. 233.
BELOUET.—Le Sanatorium de Ruppertshain, *Revue d'Hygiène et de Police Sanitaire*, 20th May, 1896, and 20th March, 1897.
—— Études sur quelques Hôpitaux Allemands. Paris, 1892.
BERNHEIM, S.—Tuberculose Pulmonaire. Paris, 1893.
BION.—Appel au Peuple Suisse en Faveur de la Création d'Hôpitaux pour les Phtisiques. Zurich, 1893.
BLUMENFELD, F.—Die Behandlung der Lungenschwindsucht in Falkenstein. Berlin, 1887.
—— Wo soll man Heilstätten für Lungenkranke errichten? *Zeitschr. f. Krankenpfl.*, 1896, No. 5.
—— Ueber den Einfluss meteorologischen Vorgänge auf den Verlauf der Bacillären Lungenschwindsucht. Inaugural Dissertation. Osnabruck, 1892; Würzburg, 1892; and Berlin, 1893.
BOURCART, A., and VIVANT, J. E.—The Importance of Climate in the Treatment of Pulmonary Tuberculosis, *Brit. Med. Journ.*, 1st Oct., 1898.
BOWDITCH, V. Y.—The Treatment of Phthisis in Sanatoria near our Homes, *Med. C. Mass. Med. Soc.*, Boston, 1896, xvii., No. 1.
—— Sanatoria for Phthisis, *Trans. Amer. Climat. Assoc.* for 1896.
—— Three Years' Experience with Sanitarium Treatment of Pulmonary Diseases near Boston. Boston, 1894.
BRAUMÜLLER.—Wiener Kranken-anstatten. Vienna, 1892.
BREHMER.—Die Aetiologie der chronischen Lungenschwindsucht. Berlin, 1885.
—— Die Therapie der chronischen Lungenschwindsucht. Wiesbaden, 1889.

British Medical Journal, 28th May, 1898 ; 1st June, 1895 ; 7th Aug., 1897 ;
21st May, 1898 ; 30th April, 1898.
—— A Sanatorium for Consumptives in Yorkshire, *Brit. Med. Journ.*, 5th
Nov., 1898.
British Med. Journ. Epitome, 17th Oct., 1896.
BROADBENT, Sir Wm.—The Prevention of Consumption and other Forms.
of Tuberculosis, *Lancet*, 29th Oct., 1898.
BRUNON.—Le Régime des Sanatoria. Paris Tuberculosis Congress, 1893.
—— Le Sanatorium du Vernet. Rouen, 1891.
BURDETT.—Hospitals and Charities. 1897.
BUSCH.—Görbersdorfer Heilanstalt für Lungenkranke. Berlin, 1875.
BURTON-FANNING, F. W.—The Open-air Treatment of Phthisis in Great.
Britain, *Practitioner*, June, 1898.
—— The Open-air Treatment of Phthisis in England, *Lancet*, 5th, 12th,.
26th Mar., 1898.
BRYN, H.—Importance of Sanitaria for the Treatment and Prevention of
Tuberculosis : a suggested locality for New York, *Sanitarian*, New
York, 1898, xl. 491.

CAVERHILL, T. F. S.—The Open-air Treatment of Consumption, with
Exhibition of Patients treated in German Sanatoria, *Ann. Mtg.
Brit. Med. Assoc.*, 1898.
—— The Value of Senatoria and the Need for their Establishment in
Great Britain, *Brit. Med. Journ.*, 1st Oct., 1898.
CALWELL, Wm.—The Hygienic Treatment of Consumption Independently
of Sanatoria, *Brit. Med. Journ.*, 1st Oct., 1898.
CHARCOT and DEBOVE.—Traité de Médecine (art. Phtisie Pulmonaire).
CHURCHILL, Fred.—Open-air Treatment in Seaside Verandahs, *Brit. Med.
Journ.*, 1st Oct., 1898.
CORMAK.—Die Natur und Behandlung der Lungenschwindsucht. Erlan-
gen, 1858.
CORNET, P.—Die Hygiene der Curorte und Heilanstalten für Tuberculose,
Buda-Pesth *International Congress of Hygiene and Demography*, 1894.
—— Prophylaxis der Tuberculose, *Berl. Kl. Woch.*, 1889.
CORNIL.—*See* Hérard.
COURTOIS-SUFFIT and BOULAY.—Traitement de la Tuberculose par
l'Aeration continue, *Gaz. des Hôp.*, 1893.

DAREMBERG.—Traitement de la Phtisie Pulmonaire—le Sanatorium de
Gausdal en Norwège. Paris, 1892.
—— Les Établissements fermés pour le Traitement des Tuberculeux.
Paris, 1892.
—— Traitement de la Phtisie Pulmonaire. Paris, 1892.
DEBOVE.—Leçons sur la Phtisie. 1883.
DETTWEILER, P.—Taschenfläschen für Hustende, *Therap. Monatsh.*, 1889.
—— Ueber die Hygiene der Schwindsüchtigen in Geschlossenen Heilan-
stalten, *Congr. for Hyg. and Demogr.*, Buda-Pesth, 1894.
—— Zur Kenntniss und Heilung der Lungenschwindsucht. Frankfort,
1886.
—— Mittheilungen über die erste deutsche Volksheilstätte für unbe-
mittelte Kranke in Falkenstein i/T., *Deutsche Med. Wochenschr.*,.
1892, 48.
—— Die Therapie der Phtisis, *6th Congr. Germ. Natural. and Med. Men*,.
Wiesbaden, 1887.
—— Die Behandlung der Lungenschwindsucht in Geschlossenen Heilan-
stalten, 2nd Ed. Berlin, 1884.
—— Bericht ueber 72 seit 3-9 Jahren in Falkenstein geheilte Fälle von
Lungenschwindsucht. Frankfort, 1886.
—— Therapie der Lungenschwindsucht. Berlin, 1884.

DETTWEILER and REBLAUD.—Traitement Hygiénique de la Phtisie Pulmonaire. Paris, 1888.

DEVIS.—Zur Therapie der Phtisis, *St. Pet. Med. Woch.*, 1894, No. 44.

DOBELL.—Bacillary Consumption. London, 1889.

DRIVER, C.—Denkschrift. 1890.

—— Volkssanatorien für Lungenkranke in *Deutsche Medizinal Zeitung*, 1890.

—— Volksheilstätten für Lungenkranke. Auerbach, 1893.

DUFOUR.—Les Sanatoria de Davos Platz et de Davos Dörfli. (*Rapport à la Société anonyme des Sanatoria Algériens.*)

DYRENFURTH.—Ueber Heilstätten für Schwindsüchtige. Berlin, 1890.

EHRENBERG, W.—Aufgabe und Ziele einer methodischen Behandlung von Genesenden in Kurorten und Heilstätten, *Vers. des Bergischen Vereins für Gemeinwohl.*, Barmen, 1894.

FRANKFORT.—Jahresbericht des Frankfurter Vereins für Reconvalescenten Anstalten. 1897.

FRÉMY.—Les Établissements fermés pour le Traitement des Phtisiques. Paris Tuberculosis Congress, 1889.

FREUPENTHAL, W.—The proposed City Hospital for Consumption, *N.Y. Med. Journ.*, 1897, lxv.

FELKIN, R. W.—Lauterberg and St. Andreasberg, in *Prov. Med. Journ.*, July, 1892.

FINKELBURG.—Ueber die Errichtung von Volkssanatorien für Lungenschwindsüchtige, *Niederrh. Verein für öffentl. Gesundheitspflege*, 1890, ix.

FISK, Sam.—Sanatoria for Phthisis, *Trans. Amer. Climat. Assoc.*, 1896.

FOSTER, Michael G.—The Mediterranean Littoral as a Health Resort for Phthisis, *Practitioner*, June, 1898.

—— See Weber, Hermann.

FOX, Wilson.—Diseases of the Lungs and Pleura. London, 1891.

FRÄNKEL, B.—Der Berlin-Brandenburger Heilstätten-verein für Lungenkranke und seine Heilstätte in Belzig, *Berl. Kl. Wochenschr.*, 1898, No. 46.

GABRILOWITSCH.—Ueber die Kaiserlichen Sanatoria für Lungenkranke zu Halila in Finnland, *Deutsche Med. Wochenschr.*, 1897, No. 9.

GANDY.—Les Sanatoria de France, *Gaz. des Eaux*, 11th Feb., 1897.

GEBHARD and HAMPE.—Die Erbauung von Heilstätten für Lungenkranke durch Invaliditäts und Altersversicherungsanstalten, Krankenkassen und Communalverbände, *XX. Vers. des deutschen Vereines für öffentl. Gesundheitspflege*, Stuttgart, 1895.

GRANCHER.—Traitement de la Tuberculose, *Bull. Méd.*, 1895 and 1896.

—— Sanatoria for Open-air Treatment, *Gaz. Méd.*, 1878.

GREY, Harry.—The Open-air Treatment of Phthisis, *Brit. Med. Journ.*, 10th Sept., 1898.

GUILLEMARD, B. J.—The Open-air Treatment of Phthisis in South Africa, *Brit. Med. Journ.*, 1st Oct., 1898.

GUY-HINSDALE.—Recent Measures for the Prevention and Treatment of Tuberculosis, *Trans. Amer. Climat. Assoc.*, 1895, vol. xi.

HÄGLER, A.—Ueber die Errichtung von Heilstätten für unbemittelte Lungenschwindsüchtige in der Schweiz, *Corr. Bl. für Schweizer-Aerzte*, xxiii. 14.

HANCE, I. H.—*New York Medical Record*, 28th Dec., 1895.

HANOT.—See Hérard.

HARE.—System of Practical Therapeutics. 1897. *Art.* by S. E. Solly.

HARRIS and BEALE.—The Treatment of Pulmonary Consumption. London, 1895.

HEILIGENSCHWENDI.—Die Bernische Heilstätte für Tuberculöse zu Heiligenschwendi.

Heilstätten Korrespondenz. Berlin—various numbers.

HÉRARD, CORNIL and HANOT.—La Phtisie Pulmonaire. Paris, 1888.

HERON, G. A.—*Lancet*, 6th Jan., 1894.

HESS, Karl.—The Treatment of Phthisis at the Falkenstein Sanatorium on Mount Taunus, *Practitioner*, Nov., 1897.

—— Heilstätte für unbemittelte Lungenkranke. Falkenstein, 1892.

HOARE, J. H.—Evidence as to American Sanatoria, *New York Med. Record*, 25th Dec., 1895.

HOHE.—Die Bekämpfung und Heilung der Lungenschwindsucht. Munich, 1897.

HOSERGEY, D.—Sanatoria for Phthisis, *Trans. Amer. Climatol. Assoc.* for 1896.

HUGHES.—Die Athmungsgymnastik bei der Lungentuberculose, *Blatt. für Klin. Hydrotherapie*, 1894, 8.

JACCOUD.—Curabilité de la Phtisie Pulmonaire. Paris, 1881.

JACOBI, E.—Das Sanatorium Arosa.

JACOBY.—Die künstliche und natürliche Hyperaemie der Lungenspitzen gegen Tuberculose durch Thermotherapie und Autotransfusion, nebst Mittheilungen aus der Heilstätte Ruppertshain im Taunus, *Münch. Med. Wochenschr.*, No. 8, 23rd Feb., 1897.

JACUBASCH.—Lungenschwindsucht und Höhenklima. Stuttgart, 1887.

—— Zur Statistik der Tuberkulose, *Prager Med. Wochenschrift*, 1892, No. 29.

—— Ueber Inhalationen, *Deut. Med. Wochenschrift*, 1889, No. 27.

—— Ueber die klimatische Behandl. der Tuberk., *Harzer Kurblatt*, 1890, No. 10.

JAHRESBERICHT.—*See* Annual Reports.

JAMES, Alex.—Pulmonary Phthisis. Pentland.

JARUNTOWSKY, Arth. v.—The Private Sanatoria for Consumptives and the Treatment adopted within them. Trans. E. Clifford Beale. London, Rebman Pub. Co.

JASINSKI.—Görbersdorf und seine Heilanstalten, *Petersb. Med. Wochenschrift*, 1887.

JOHNS, W. H. Denton.—The Open-air Treatment in England, *Brit. Med. Journ.*, 1st Oct., 1898.

JONES, P. Sydney.—Cases suitable for the Open-air Treatment in Australia, *Brit. Med. Journ.*, 1st Oct., 1898.

JULIUSBERGER.—Die Sanatorien. Berlin, 1890.

KAATZER.—Bad Rehburg: eine Heilstätte für Lungenkranke. Hanover, 1885.

KAURIN, Ed.—Rapport sur un Voyage d'Étude aux Établissements de Tuberculeux d'Allemagne et de la Suisse, *Norsk. Mag. f. Lagevidensk.* Christiania.

KLEBS, A. C.—The Necessity of Special Institutions for the Treatment of Pulmonary Tuberculosis, in the *Tri-state Med. Journ. and Pract.*, St. Louis, May, 1897.

KNOPF, S. A.—The Hygienic Educational and Symptomatic Treatment of Pulmonary Tuberculosis, with a plan for Sanatoriums for the Poor, *N.Y. Med. Record*, 1897, li. 222.

—— The Present Status of Preventative Means against the Spread of Tuberculosis in the various States of the Union, critically reviewed, *J. Amer. Med. Assoc.*, 30th Oct., 1897.

—— Les Sanatoria pour la Phtisie Pulmonaire. Paris, 1895.

—— The Urgent Need of Sanatoriums for the Consumptive Poor of our large Cities, *N.Y. Med. Rec.*, 1897, 3rd Oct., 1896.

KOBERT, Prof. Dr.—Görbersdorfer Veröffentlichungen, No. 1. Stuttgart, 1898.

KÖLBL, J.—Die hyg. diätet. oder abhärtende Behandlungen der Lungenschwindsucht, *Wiener Med. Presse*, xxxviii. 1573-1577.

KORÁNYI.—Lungenschwindsucht, *Eulenburg's Real-encyclopädie*.

KÜCHLER.—Die Errichtung einer Heilstätte für die Stadt Worms. Worms, 1893.

KUTHY, D.—Sanatoria für Lungenschwindsucht, *Pester Med. Chir. Presse* 1898, xxxiv.

—— A Tüdövész Szanatóriumi Gyógyitása. Budapest, 1897.

—— Ueber Lungenheilanstalten. Leipzig, 1898.

—— Tüdövészes Szanatoriumokról, *Rev. des Cliniques*, Budapest, 7th yr., 4th vol.

—— Görbersdorf und Alland, *Weekly Med. Gaz.*, Budapest, 1896, 50.

LAGRANGE.—Le Traitement Hygiénique en Allemagne, *Rev. des Maladies de la Nutrition*, 1895.

Lancet, 30th May, 1895 ; Feb., 1897.

—— English Sanatoria for Consumptives. *Lancet*, 2nd April, 1898.

—— The Open-air Treatment of Consumption (Nathan Raw at Liverpool), *Lancet*, 29th Oct., 1898.

—— The Climatic Treatment of Pulmonary Phthisis, *Lancet*, 22nd Jan., 1898.

LATULLE.—Hospitalisation des Phtisiques, *Semaine Méd.*, 4th May, 1892, and *Presse Méd.*, 11th Aug., 1894.

LEYDEN, E. v.—Ueber den gegenwärtigen Stand der Behandlung Tuberculöser und die staatliche Fürsorge für dieselben. Berlin, 1897. Also *Zeitschr. f. Krankenpfl.*, Sept. and Oct., 1897.

—— Die Versorgung Tuberculöser seitens grosser Städte, *Congr. Hyg. and Demogr.*, Budapest, 1894.

—— Ueber Pneumothorax Tuberculosus nebst Bemerkungen ueber Heilstätten für Tuberculöse, *Deut. Mediz. Wochenschr.*, 1890, No. 7.

LEYSIN.—Le Sanatorium de Leysin. Paris, 1893.

LIEBE, G.—Zwei neue Spuckflaschen für Tuberculöse, *Illustr. Monatssch. der Aerztliche Polyl.*, Berlin, 1898, xx. 92.

—— Der Stand der Bewegung der Volksheilstätten für unbemittelte Lungenkranke in Deutschland, *Hygienische Rundschau*, 1897, No. 21 ; 1895, No. 17 ; 1896, Nos. 13, 14, 16.

—— Ueber die dringende Nothwendigkeit der Errichtung von Volksheilstätten für unbemittelte Lungenkranke, *Therap. Monatsh.*, Nov., 1897.

—— Ueber Volksheilstätten für Lungenkranke. Breslau, 1895.

—— Beiträge zur Volksheilstättenfrage, *Hyg. Rundschau*, 1895, 17.

—— Die Behandlung de Lungentuberculose in Volksheilstätten mit bes. Beziehung auf die Volksheilstätten-methode v. Rothen Kreuz Grabowsee, *Deutsche Militär-aerztliche Zeitschr.*, Berlin, 1897, xxvi. 471.

—— Ziele und Wege zur Bekämpfung der Tuberculose, *Therap. Monatsh.*, Berlin, 1897, xi. 577.

LINDSAY, J. A.—The Climatic Treatment of Consumption. London, 1887.

—— The Problem of the Consumptive Poor, *Lancet*, 4th Dec., 1897.

LOHMANN, W.—Die Gründung der Heilstätten für unbemittelte Lungenkranke. Hanover, 1890.

LOOMIS.—The Training School for Nurses at the Loomis Sanitarium, *Lancet*, 2nd April, 1898.

LYMAN, H.—Dietetic and Hygienic Treatment of Consumption in Colorado, *Diet. and Hyg. Gaz.*, N.Y., 1897, xiii. 487.

MASZÁK, E.—A Tuberculosis Gyogyitása és a Szanatoriumok.
MAYER.—Die Tuberculose : Behandlung in Sanatorien. Vienna, 1893.
MEISSEN, E.—Das Sanatorium Hohenhonnef am Rhein, *Verein der Aerzte des Bezirks Köln geh.*, 9th May, 1896, Bonn, 1896.
—— Ueber eine neue Heilanstalt für Lungenkranke, *Centralblatt für allgemeine Gesundheitspflege*, 1889.
—— Fieber und Hyperthermie, *Berl. Kl. Wochenschrift*, 1898, No. 23.
—— Was können die Fachärzte zunächst zur Bekämpfung der Tuberculose thun ? *Therap. Monatsh.*, 1897, No. 11.
MACCORMAC.—Air re-breathed.
MANASSE.—Die Heilung der Lungentuberculose durch diätetisch-hygienische Behandlung in Anstalten und Kurorten. Berlin, 1891.
MANNHEIMER.—Die hyg. diät. Behandlungsmethode der Lungentuberculose nebst Bemerkungen ueber die Versorgung Tuberculöser Armen, *N.Y. Med. Wochenschr.*, 1897, ix.
MARFAN.—Une Visite au Sanatorium du Canigou, *Gaz. des Hôp.*, 17th Sept., 1891.
—— *See* Charcot and Debove.
MARTINEAU, E. Marty.—The Sanatorium of Mont Bonmorin, *Indépendance Médicale*, 25th Mar., 1896.
MICHAELIS (Rehburg).—*Monatsheft f. prakt. Balneol.*, 1897, No. 1.
MÖLLER.—Les Sanatoria pour le Traitement de la Phtisie Pulmonaire. Bruxelles, 1894.
—— Les Sanatoria pour les Phtisiques Pauvres, *Mouvement Hygiénique*, April, 1893.
—— Le Sanatorium de Falkenstein pour les Phtisiques Pauvres. Brussels, 1893.
MONTMEYLIAN, M. E.—Un Hivernage dans un Sanatorium Alpin. Paris, 1896.
MORITZ.—Sanatoria für Lungenkranke. Brunswick, 1892.
—— Sanatorien für Lungenkranke, *XVII. Vers. des deutschen Vereines für Gesundheitspflege*, Leipzig, 17th Sept., 1891 ; *Deutsche Vierteljahrsschr. für öffentl. Gesundheitspfl.*, xxiv. 4, 1.
MÖRSÜL.—Sanatorium Mörsül. *Hälsovännen*, Stockholm, 1897, xii.

NAHM.—Die neue Heilstätte für unbemittelte Lungenkranke zu Rupportshain im Taunus, *Zeitschr. f. Krankenpfl.*, 1896, 2.
—— Ueber die Einrichtung von Volksheilstätten, *Zeitschr. für Krankenpfl.*, 1898, Nos. 6 and 7.
—— Sind Heilstätten eine Gefahr für ihre Umgebung ? *Münch. Med. Wochenschrift*, 1895, No. 40.
NETTER and BEAULAVON.—Du Traitement des Tuberculeux Indigents dans les Sanitoriums, *Gaz. Hebd. de Med. et de Chir.*, No. 66, 18th Aug., 1898 ; and *Presse Méd.*, Paris, 1898, ii.
NICAISE.—De l'Établissement d'un Sanatorium pour les Phtisiques.
NICKSTEADT.—*Reichsmediz. Anzeiger*, 1896, No. 6.

OEHME.—Ueber Volksheilstätten und ueber das Sanatorium des Sächsischen Landesvereines für unbemittelte Kranke in Reiboldsgrün. Auerbach, 1896.
ÓNODI, A.—Tüdőbajosok Sanatoriumok és klimatikus helyeken, *Órvosi Hetilap.*, Budapest, 1898, xlii. 222.
OTIS, E. A.—Some Modern Methods in the Treatment of Phthisis and its Symptoms, *Boston Med. and Surg. Journ.*, 14th and 21st July, 1898.
—— Hospitals and Sanatoria for Consumptives Abroad, *Boston Med. and Surg. Journ.*, cxxxviii. 265, 313.
—— Sanatoria for Phthisis, *Trans. Amer. Climat. Assoc.* for 1896.

PALLESKE.—Görbersdorf in Schlesien : eine Heilanstalt für Lungen-kranke. Berlin, 1892.
PAYNE, Thos. H. Khyber.—Instructions as to Washing Patients' Pocket-handkerchiefs. Ventnor, 1893.
PENZOLDT.—Behandlung der Lungentuberculose, in *Handbuch der Special-len Therapie der inneren Krankheiten.*
PETIT, E. P. Léon.—Le Phtisique et son Traitement Hygiénique. Paris, 1895.
—— L'Hôpital de Villiers : l'Œuvre des Enfants Tuberculeux. Paris, 1896,
—— *Proc. Tuberculosis Congress,* Paris, 1898. See *Médecine Moderne,* 3rd Aug., 1898, No. 60.
—— *Rev. de la Tuberculose,* April, 1896.
PHILIP, R. W.—Remarks on the Universal Applicability of the Open-air Treatment of Pulmonary Tuberculosis, *Brit. Med. Journ.,* 23rd July, 1898.
PLIQUE, A. P.—Le Sanatorium d'Angicourt, in De la Harpe's *Les Stations d'Hiver.*
POLLOCK, J. E.—The Hospital Treatment of Consumption, in the *Prac-titioner,* June, 1898.
—— Elements of Prognosis in Consumption. London.
POORE, G. Vivian.—The Climate of the Dwelling-house, *Journ. of Balneol. and Climatol.,* Oct., 1897.
POTT, Fras.—The Open-air Treatment of Phthisis in England, *Lancet,* 2nd April, 1898.
POUZET, Paul.—Les Sanatoria de Görbersdorf, *Progrès Médical,* 1st Nov., 1890.
POWELL, Sir R. Douglas.—Diseases of the Lungs, 4th Ed. London, 1893.
Practitioner, The, June, 1898.
PREYSZ, K.—A Tuberculosis a közegészség pontjábol s annak gyógyitása tekintettel hazai fürdönkre.

RANSOME, A.—*Lancet,* 11th Jan., 1898.
—— Some Results of Open-air Treatment of Phthisis at Bournemouth, *Lancet,* 11th June, 1898.
—— The Treatment of Phthisis. London, 1896.
—— Researches on Tuberculosis, Weber-Parkes Prize Essay. London, 1898.
RANSOME and DRESCHFELD.—*Pr. Royal Soc.,* xlix.
RANSOME and DELÉPINE.—*Pr. Royal Soc.,* lvi., May, 1894.
REINHARDT, C.—*Brit. Med. Journ.,* 7th Aug., 1897.
RICHTER, C. M.—The Formation in California of Colonies for Consump-tives, *Pr. Med. Soc. Calif.,* San Francisco, 1897, 68.
ROBINSON, Wm.—On the Repression of Consumption. Pres. Addr. to the N. of Engl. Branch of the British Medical Association, 11th July, 1898, *Brit. Med. Journ.,* 23rd July, 1898.
ROBERTS, J. Lloyd.—The North Wales Coast as a Health Resort and for the Open-air Treatment of Phthisis, *Med. Magazine,* Nov., 1898, vol. vii., No. 11, p. 869.
RÖMPLER, H.—Die Behandlung Lungenkranker in Höhenkurorten, mit specieller Berücksichtigung der geschlossenen Heilanstalten. Buda-pest, Preyszsche Bibliothek, No. 36.
—— Prophylaxis gegen Lungenschwindsucht in den Kurorten, *Monatsschr. f. prakt. Balneol.*
—— Zur Geschichte Görbersdorf.
—— Die Frage der Kontagiosität der Tuberculose, *Deutsche Medizinal Zeitung,* 1898, No. 35.
—— Beiträge zur Lehre von der chronischen Lungenschwindsucht. Ber-lin, 1892.

Rosin, H.—Die englischen Schwindsuchts-Hospitäler und ihre Bedeutung für die deutsche Schwindsuchtspflege, *Deutsches Vierteljahrschr. f. öffentl. Gesundheitspfl.*, xxiv., 1892.

Rothe Kreuz, Das.—Various numbers.

Ruck, K. v.—Chronic Pulmonary Tuberculosis, *Virginia Med. Semi-Monthly*, Richmond, 1897-8, ii. 217.

Ruppertshain.—*Zeitschr. f. Krankenpfl.*, 1896, No. 2; *Rev. de la Tuberculose*, Dec., 1896.

Sabourin, Ch.—Traitement Rationnel de le Phtisie. Paris, 1896.

—— *Gaz. Hebdomadaire*, 31st October, 1891.

Sandwith, F. M.—Desert Climates for Lung Tuberculosis, *Practitioner*, June, 1898.

Schill and Fischer.—*Mitth. a. d. K. Gesundheitsamt*, Bd. ii., 1884.

Schmey, F.—Zur Behandlung der Tuberculose, *Aerztl. Central Zeitung,.* Vienna, 1898, x. 113.

Schmid.—Ueber Volkssanatorien für Lungenkranke, *Münch. Med. Wochenschr.*, 1898, Nos. 11 and 12.

Schrötter, Prof.—Die Heilanstalt in Alland, *Fourth Annual Report,.* Vienna, 1896.

Schultze.—Die Therapie der Lungenschwindsucht, *Kalender für Aerzte*, 1879.

Schultzen.—Die Behandlung der Lungentuberculose in Volksheilstätten, *Deutsche aerztliche Zeitschr.*, Berlin, 1897, xxvi. 471.

Schütze, C. C.—Die Hydrotherapie der Lungenschwindsucht, *Arch. der Baln. und Hydrother.*, Halle a/S, 1897, 4 and 5.

Sersiron, G.—L'Initiative Privée et les Sanatoriums pour Tuberculeux Adultes et Pauvres, *Presse Méd.*, Paris. 1898, ii. 63.

Silverskiöld, P.—Two Hospital Institutions in Göteborg to combat Tuberculosis, *Halsovännen*, Stockholm. 1897, xii. 7-21.

Smyth, R. Mander.—Remarks on the Rational Treatment of Phthisis, with reference to Nordrach Sanatorium, *Brit. Med. Journ.*, 1st Oct., 1898..

Solly, S. E.—Medical Climatology. London, 1897.

—— The Present Treatment of Tuberculosis, *Hare's System of Pract. Therap.*, 1897, iv.

Sonderegger.—Heilstätten für Brustkranke in der Schweiz. St. Gall, 1894.

Squire, J. E.—The Hygienic Prevention of Consumption. London, 1893.

Stubbert, J. E.—Sanitarium Treatment of Pulmonary Tuberculosis, *N.Y.. Med. J.*, 1898, lxviii. 159.

—— Methods of Treatment at the Loomis Sanitarium for Consumption,. *Phila. Med. J.*, 1898, 467.

—— Compar. Diagnosis by the Röntgen Rays, *Med. Rec.*, 22nd June,. 1897.

Székely, A.—Gyógyintézetek Szegény Tüdővészesek szamára.

Szontagh, Von.—Die klimatischen Verhältnisse von Bad Neu-Schmecks.. 1894.

—— Lungenkranke in der subalpinen Region in Neu-Schmecks. 1884.

Thorne, W. Bezly.—The Open-air Treatment in London, *Brit. Med.. Journ.*, 1st Oct., 1898.

—— Lecture on Falkenstein to the British Nurses' Association.

Thorowgood.—The Climatic Treatment of Consumption. 1868.

Trudeau, E. L.—Le Sanatorium d'Adirondack. New York, 1894.

—— Sanatoria for the Treatment of Incipient Tuberculosis, *N. Y. Med. Journ.*, 1897, lxv. 276.

Turban, K.—Normalien für die Erstellung von Heilstätten für Lungenkranke in der Schweiz.

—— Die Heilanstalt für Lungenkranke, Davos.

UNTERBERGER, S. A.—Ueber die Nothwendigkeit der Einrichtung von
    Haus-Sanatorien für Tuberculöse in den Militär-hospitälen mit
    Berücksichtigung des heutigen Standpunktes der Tuberculose,.
    Deutsche Militär-aerztliche Zeitschrift, 1897, xxvii.
—— Lungentuberculose und ihre Behandlung speciell in Haus-Sanatorien,
    St. Pet. Med. Woch., 1896, n. F. xiii. 29, 32, and 1897, n. F. xiv. 2;.
    also in *Blätter der Kl. Hydrotherapie*, Nov., 1897.

VENTNOR.—Description and Views of the Royal National Hospital for·
    Consumption and Diseases of the Chest, on the separate principle,
    situated in the Undercliff, Ventnor, I. W. 1895.
VOLLAND, A.—Die Behandlung der Lungenschwindsucht im Hochgebirge..
    Leipzig, 1889.
—— Die Lungenschwindsucht und ihre Entstehung, Verhütung, Behand-
    lung, und Heilung. Tübingen, 1898.

WALKER, Miss Jane.—Experience of the Open-air Treatment in England,.
    Brit. Med. Journ., 1st Oct., 1898.
WALSHE.—Diseases of the Lungs, 4th Ed.
WALTERS, F. Rufenacht.—Observations on Sanatoria for Consumptives,
    Medical Magazine, Nov., 1898, vol. vii., No. 11.
—— The.Prevention of Tuberculosis, *Lancet*, 1898, ii. 258.
—— Sanatoria for Consumptives, *Practitioner*, June, 1898.
—— On the Climatic Treatment of Phthisis, *Lancet*, 20th Nov., 1897.
—— Some German Sanatoria for Phthisis, *Lancet*, 14th Aug., 1897.
WEBER, Hermann.—The Hygienic and Climatic Treatment of Chronic
    Pulmonary Phthisis (Croonian Lecture). 1885.
WEBER, Hermann, and Michael G. FOSTER.—Climate in the Treatment
    of Disease, in *Clifford Allbutt's Syst. of Medicine*, vol. i.
WEBER, F. Parkes.—Ocean Voyages in Phthisis, *Practitioner*, June, 1898.
WEICKER, J.—Beiträge zur Frage der Volksheilstätten, *Zeitsch. f. Kran-
    kenpl.*, 1896, Nos. 3 and 4.
—— Die Volksheilstätten für Lungenkranke in ihrer social-politischen·
    Bedeutung, *Köln. Ztg.*, 8th Dec., 1895.
—— Beiträge zur Frage der Volksheilstätten. Friedland, 1897.
WEISSMAYR, Alex. Ritter v.—Die Furcht vor Heilstätten für Tuberculöse,
    Wiener Kl. Woch., 1897, x., and *Zeitschr. f. Krankenpfl.*, Dec., 1897.
—— Die Schweizer Volksheilstätten für Tuberculose, *Wiener Kl. Woch.*,.
    1897, 2.
WIJNHOFF.—Sanatoria voor Phthisici, *Nederl. Tijdschr. v. Geneesk.*
    Amsterdam, 1897, 2 R. xxxiii.
WILKS, Sir Saml.—The Treatment of Consumption, *Practitioner*, June,
    1898.
WILLIAMS, C. Theodore.—The Treatment of Pulmonary Tuberculosis by ᵥ
    Residence at High Altitudes, *Practitioner*, June, 1898. ᵥ
—— *Journal of Balneol. and Climatol.*
—— A Lecture on the Open-air Treatment of Pulmonary Tuberculosis as
    practised in German Sanatoria, *Brit. Med. J.*, 1898, i. 1309. ⌐
—— Pulmonary Consumption. ᵥ
—— Aerotherapeutics. ⌐
WISE, Tucker.—English Sanatoria for the Treatment of Phthisis, *Lancet*,.
    2nd April, 1898.
WOLFF, J.—Die moderne Behandlung der Lungenschwindsucht.
WOLFF AND SAUGMANN.—Ueber die dauernde Heilung der Tuberculose.
    Wiesbaden, 1891.
WOLFF, F.—Heilstätte für unbemittelte Brustkranke, *Munch. Med. Woch.*,.
    1892, 51.
—— Sächsische Volksheilstätte für Lungenkranke. Auerbach, 1892.

WOLFF, F.—Ueber die Gründung einer Sächsischen Volksheilstätte in der
    Nähe von Reiboldsgrün im Vogtland.   Auerbach, 1898.
WOLFF-IMMERMANN, F.—Bericht des Vereines für Begründung und Unter-
    haltung von Volksheilstätten für Lungenkranke im Königreiche
    Sachsen.   Auerbach, 1896.

YEO, J. Burney.—Health Resorts.   London, 1893.

ZIEMSSEN, v.—Ueber den gegenwärtigen Stand der Behandlung Tubercu-
    löser.   Berlin, 1897.
ZUBIANI, Ausonia.—La Cura Razionale dei Tisici ei Sanatorii, *Lancet*, 25th
    June, 1898.

INDEX.

ACHTERMANN, Dr., 67, 150, 177, 186.
Adirondack Cottage Sanitarium, 18, 20, 23, 34, 61, 66, 77, 79, 83, *84*.
—— Mountains, 84, 87.
Administration, 59.
Admission of Patients, 1, 49, 57, 60, 62.
Aegeri Sanatorium for Children, 283, 299, 310.
Agnetz Sanatorium, 128.
Agricultural Colonies, 62, 144, 312.
Aigle, 293.
Aiken Cottages, 84, *93*.
Aix-la-Chapelle, 246.
Alabama. See Hygeia.
Albersweiler, 206.
Albertsberg, 22, 46, 75, 78, 206, *222*.
Albrecht, Archduke, 121.
Albrechtshaus Sanatorium, 206, *232*.
Alcohol, 43.
Alexander Sanatorium, Halila, 273, *277*.
Alexandre Hospital, St. Petersburg, 272-3.
Alexandrina Hospital, St. Petersburg, 272-3.
Algiers Sanatorium, 129.
Alland Sanatorium, 120.
Alpine Climates. See Climates.
—— Sanatoria. See Sanatoria.
Alsace-Lorraine, 206.
—— Insurance Co., 260.
Alteua Sanatorium, 75, 77, 206, 246, *248*.
—— Johanniter Isoliranstalt, 78, 251.
Alteubrak Sanatoria, 148-9, *167*, 206, *239*.
Alteration of Existing Buildings, 73.
Altitude, 5, 13, 14, 54.
Altona Sanatorium, 75, 246.
American Homes. See Homes.
—— Hospitals. See Hospitals.
—— Sanatoria. See Sanatoria.
Angicourt Sanatoria, 128.
Anguey, Dr., 97.
Arcachon, 128.
Ardennes, Belgian, 311.
Arles, 136.
Area Required, 16, 73.
Arlen Sanatorium, 78, 206, *201*.
Arosa Sanatorium, 79, 283, *284*.
Asheville. See Winyah.
Aspect, 13, 21.

Assistance Publique, 80, 127-8.
Assuan (Egypt), 312.
Atmospheric Purity, 14.
Aufrecht, Dr., 65.
Aussee (Styria), Dr. Schreiber's Sanatorium, 126.
Austin Hospital, Australia, 315.
Australia, 315.
Austria-Hungary, Sanatoria in, 120.
Avants, Les, 283.

BACILLI, Tubercle, 28.
Baden, 148, 190, 199, 201, 206, 261-2.
—— (Vienna), 121.
—— Insurance Co., 257, 262.
Badenweiler Sanatorium, 190, *201*.
Badische Anilin und Soda Fabrik, 257.
Bailey, Dr. W. C., 108.
Balconies, 19.
Baltic Coast, 207, 214, 277, 280.
Baracken, Döckersche, 73, 217, 234.
Bariatinsky, Countess, 273.
Basel Sanatorium. See Davos.
—— Gemeinnützige Gesellschaft, 299.
—— Hilfsverein für Lungenkranke, 302.
Bathrooms, 20.
Baths, 36.
Batthyaui, Count, 120.
Battle Creek Sanitarium, 109.
Baudach, Dr., 197.
Bavaria, 206, 263, 265.
Beaulavon, Dr., 52, 112.
Beds, 45, 58, 88, 349.
Beelitz Sanatorium, 206, *214*.
Belfast Hospital for Consumption, 320, 333.
Belgium, 311.
Bellevue Hospital, 66.
—— Sanitarium, 84, *113*.
Belzig Sanatorium, 76, 206, 214, *215*.
Beneke, Dr., 13.
Bentley, Dr., 312.
Berck-sur-Mer, 128-9.
Berka, Bad, 206.
—— Sanatorium, 206, *244*.
Berlin, 216, 219.
—— Brandenburg Sanatorium Society, 208-4, 214-15.
—— Insurance Co., 214.

24

ABERDEEN UNIVERSITY PRESS.